Design and Analysis of Spiral Inductors

Genemala Haobijam · Roy Paily Palathinkal

Design and Analysis of Spiral Inductors

Springer

Genemala Haobijam
R&D 3, ATG, SEL
Samsung Noida Mobile Centre
Noida
India

Roy Paily Palathinkal
Department of Electronics and Electrical Engineering
Indian Institute of Technology Guwahati
Guwahati, Assam
India

ISBN 978-81-322-2904-9 ISBN 978-81-322-1515-8 (eBook)
DOI 10.1007/978-81-322-1515-8
Springer New Delhi Heidelberg New York Dordrecht London

Printed on acid-free paper

Springer is part of Springer Science+Business Media (www.springer.com)

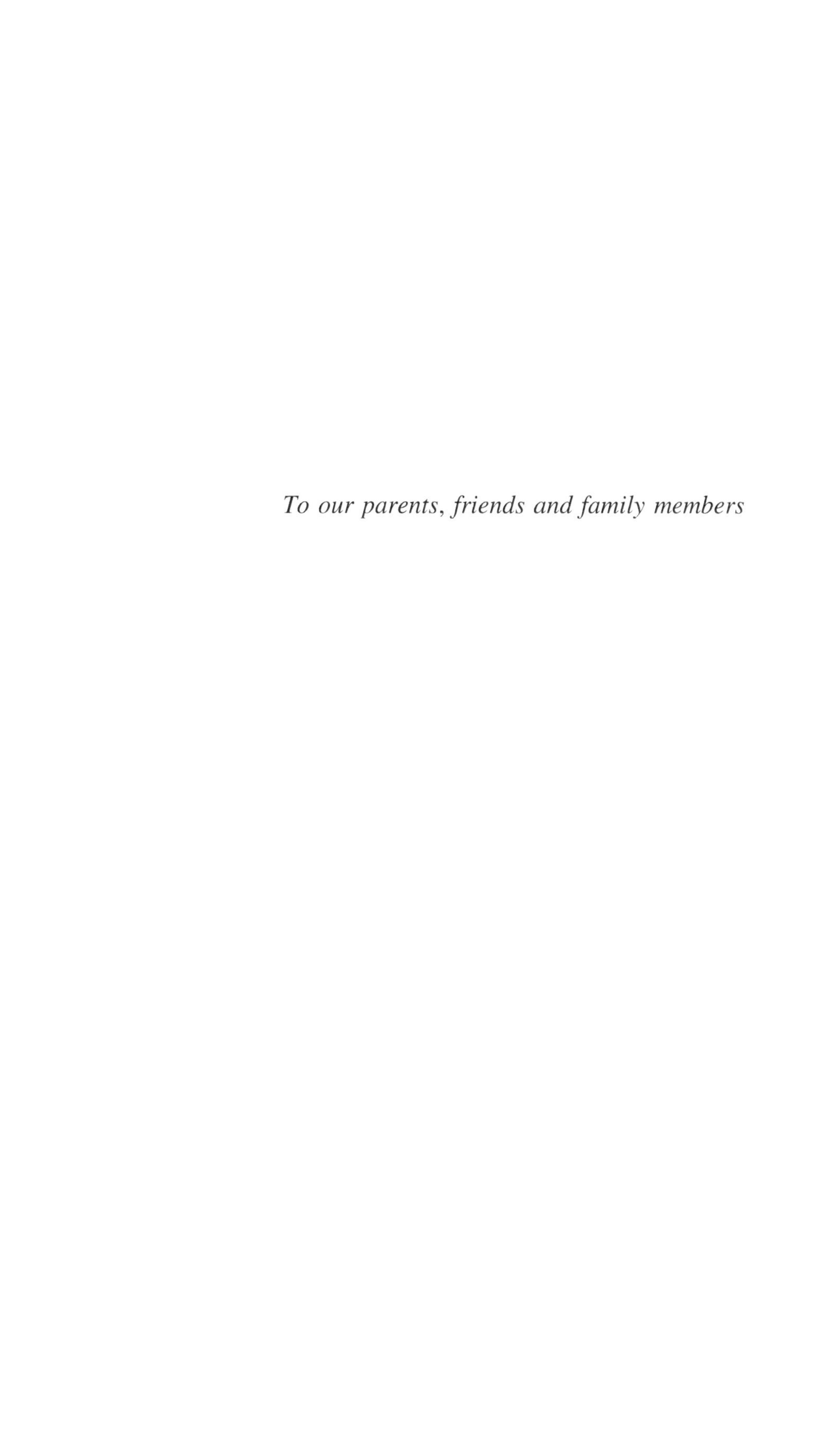

To our parents, friends and family members

Preface

The ever expanding wireless and consumer electronics market necessitates the integration of more and more multiple functions and there is a growing demand for small size, low cost, and high performance circuits. Interestingly today, many wireless applications necessitate the integration of many more multiple functions like phone, video-game console, personal digital assistant, digital camera, web-browser, e-mail, etc. This has presented a challenge to integrate the analog, digital, and radio frequency systems on a single chip. One of the biggest hurdles in the realization of SoC is the integration of the passive components, especially in RF systems. The need for integration of more functionalities has changed the direction in the development of passive components in the last several years. Of the passive devices, the most critical is the inductor.

The performance of CMOS RFICs such as voltage controlled oscillators (VCOs), low noise amplifiers (LNAs), passive element filters, etc., are well determined by the quality of inductors. For example, the quality factor of the inductor determines the stability and phase-noise power of an oscillator for all communication applications and the ability to implement extremely selective filters with a small percentage bandwidth, small shape factor, and low insertion loss. Hence, the design of an inductor is one of the critical steps in the design cycle since the performance and cost will depend on the quality factor and area of the inductor.

The figure of merits of on-chip spiral inductors is determined by their geometrical or layout and the technological parameters. There exist numerous tradeoffs between the performance of a spiral inductor and its design parameters. In most of the performance trend studies reported in the literature, the layout parameters were systematically varied and the corresponding changes in the inductance, quality factor, and resonance frequency were reported. This approach is useful in applications where one has the flexibility to choose from a range of inductance values. However, if a designer targets designing a specified inductance value and optimizes its layout parameters for a particular application, such studies give insufficient information. This is because the quality factor and the inductance follow an opposite trend with the layout parameters. For example, one may attempt to increase the quality factor by increasing the inner diameter to minimize the eddy current effect which however alters the value of inductance. In turn, one can then vary the number of turns, metal width, and spacing to get back the desired inductance value. However, this approach again alters the quality factor but this need not be the optimum value at the desired frequency. Therefore, in this book, the importance of studying the performance trend by varying the layout parameters keeping the inductance value constant is discussed in detail.

Inductors are generally designed either based on a library of previously available fabricated inductors or using an electromagnetic simulator. The former method limits the design space and the latter is computationally expensive and time consuming. A typical spiral inductor design problem is to determine the layout parameters that results in the desired inductance value. For a desired inductance value, a number of possible combinations of these parameters exist. Therefore, it is imperative to find the optimized parameters for a particular inductance that results the highest Q at desired frequency. In this book, a simple method to determine the optimum layout parameters is described which can shorten the design and product time-to-market cycle.

Generally, inductors may be optimized using enumeration or numerical methods. The enumeration methods are simple and can find a nearly global optimum design but highly inefficient. On the other hand, numerical methods have proven to be more efficient in reducing the computation time by converging rapidly to the optimal design. However, such algorithms result in a single set of inductor design parameters and no information is available to the extent the other combinations are from the optimal one. Information about the near optimal solution is also important to judiciously explore the tradeoff between the different competing figure of merits. The efficiency and the result of optimizing all such methods requires the knowledge of performance trends of the inductor with its layout parameters in order to decide the design search space. If the designer is not well acquainted with the complexity of inductor design, the design parameter constraints may include sets of unfeasible specifications which will increase the number of function evaluations and computation time unnecessarily. For example, a large search space may be defined which will require huge computation time or a small search space may be defined where the solution may not be globally optimum. Furthermore, this book discusses the methodology to find the bounds of optimization constraints and restricts the search to the feasible region and promotes fast convergence to a solution. The incorporation of a bounding method with an optimization schedule will definitely speed up the optimum inductor synthesis.

Inductors are also designed using an electromagnetic simulator. This method is computationally expensive and time consuming due to which design methods based on lumped element model are generally adopted. A lumped element model however gives only an approximate electrical characteristic and the result may also be prone to errors. Verification of the design using a full wave EM simulator is therefore required before fabrication. Sometimes the designer may even be compelled to repeat the entire design when such errors are not tolerable. Therefore, optimization using an EM simulator would be more advantageous. But a method using an EM simulator would be acceptable only if a few structures have to be simulated. In this book, we have shown that this can be made possible by identifying the optimum width and the number of turns from the simulation of a few structures. If these few structures can be identified, then the optimized design parameters can be determined most accurately using an EM simulator.

In most of the integrated circuits like amplifiers, mixers, oscillators, etc., the differential topology is preferred because of its less sensitivity to noise and interference. For such applications, symmetric inductors are preferred because under differential excitation, quality factor and self-resonance frequency increases. Generally, a pair of asymmetric planar inductors connected together in series or the conventional symmetric inductor is used, notwithstanding the area occupied, which is very large. With technology scaling, the number of metal layers is increasing and taking advantage of this; in this book, a new multilayer inductor is discussed to improve the performance of on-chip inductors. Further, its performance is demonstrated by fabricating and characterizing the inductors. The structure is implemented in an application circuit and performance is illustrated by test and measurement results.

Chapter 1 discusses the design of on-chip inductor with a review of its innovative structure evolution and design trends, followed by a discussion of the unsolved problems.

Chapter 2 exemplifies the importance to study the performance trend more systematically, keeping the inductance constant and varying the layout parameters. This study will lead to promising conclusions that in turn would help optimizing inductors more efficiently. Also, a method of bounding the layout design parameters is proposed, thereby limiting the feasible design search space and hence optimization can be carried out efficiently. Performance characterization using an EM simulator is more accurate compared to a lumped element model. This chapter also suggests a method to identify only the few nearly optimum structures and finds the most optimized design parameters using an EM simulator.

In Chap. 3, a multilayer spiral inductor is proposed, in which the traces of the metal spiral up and down in a pyramidal manner exploits the multiple metal layers. This structure is discussed extensively with the development of a lumped element model and calculation of its parasitic capacitance to predict its self resonance frequency. It is also shown that this form of spiraling results in lower parasitics. The performance trends of this new inductor with its layout parameters are also investigated. The structure is also symmetric and it is illustrated that, for differential circuit implementations, the area of the chip can be reduced to a large extent as compared to its equivalent conventional inductors. The layout, fabrication and measurement results of the inductor are also reported in detail.

Chapter 4 discusses the design of an LC differential VCO employing the proposed inductor in the LC tank. The design process of the tank inductor and the capacitor is explained. A prototype of the VCO is implemented in 0.18 μm UMC RF CMOS technology. The performance of the VCO is investigated by simulation and validated by the testing and measurement results.

March 2013

Genemala Haobijam
Roy Paily Palathinkal

Acknowledgments

We sincerely thank Prof. A. K. Gogoi, Prof. A. Mahanta for their advice and suggestions in addition to our timeless technical discussions which have encouraged us always. We also thank the faculty of the Department of Electronics and Electrical Engineering, Indian Institute of Technology, Guwahati for its encouragement. We thank Prof. G. S. Visweswaran and Dr. S. Chatterjee of Indian Institute of Technology, Delhi, for their helpful advice on the chip tape out and Dr. Bhardwaj Amrutur, of Indian Institute of Science, Bangalore for permitting to use the testing and measurement facility at VLSI Circuits and Systems Laboratory, Indian Institute of Science, Bangalore. We thank the co-ordinators of the National Program on Smart Materials Project, Dr. P. S. Robi, Dr. A. Srinivasan, and Dr. R. Bhattacharjee. The design software Intellisuite of IntelliSense Software Corp., extensively used for the design and simulation of inductor, was procured under this project. We also thank the Department of IT, Government of India. The custom IC design tools of Cadence Design Systems and the chip fabrication were funded by Special Manpower Development Project II of the Department of IT, Government of India. The encouragement from the Convener of the SMDP project, Dr. Debashis Dutta, Group Coordinator, R&D in Electronics and, DeitY is gratefully acknowledged. We thank K. C. Narasimhamurthy, M. Sabarimalai Manikandan, and P. Krishnamoorthy for their earnest support. We also thank all the students and staff, including technical personnel, of our VLSI and Digital System Design Lab. We thank Annop Niar, Sultan Siddiqui, Hitesh Shrimali, Roohie Kaushik, and Sonika of IIT Delhi for their timely help and support during the chip layouting. We also thank Bishnu Prasad, K. Shyam, Anand Seshadri, Pratap Kumar and, M. Nandish of IISc Bangalore for their help and support during chip testing. We would like to express our heartfelt gratitude to our family and friends for their constant support, understanding, love, and encouragement. The journey of writing this book has been smooth because of them who have always boosted us up in all our hardship. Without them, it would have been impossible to complete this book. We thank all of them for understanding us always.

March 2013

Genemala Haobijam
Roy Paily Palathinkal

Contents

Chapter 1
Introduction

On-chip inductors have become increasingly important and with the increasing frequencies of operation of the circuits, the on-chip inductors has gained even more importance [1]. Complementary metal oxide semiconductor (CMOS) technology has been widely adopted for its mature and mass productivity [2, 3]. Steady improvements in the radio frequency characteristics of CMOS devices via scaling is driven by advancement in lithography. It has enabled increased integration of RF functions. Spiral inductors are widely used even at microwave frequencies and their applications in millimeter-wave circuits are investigated [4]. In this chapter a brief summary of the silicon integrated passive devices is given in Sect. 1.1. An introduction to on-chip inductor is presented in Sect. 1.2. The losses in the conductor and the substrate are also explained. An overview of the evolution and progress of the on-chip inductor with a review on the integrated inductor design is presented in Sect. 1.3. The design complexity and performance issues are also discussed.

1.1 Silicon Integrated Passive Devices

In any typical printed circuit board, the component count of passive devices usually dominates that of the active devices. Surface mount passive devices were used earlier and even today in many applications they are still in use. In a cell phone board, the passive devices used to account for 95 % of the total component count, 80 % of the board area and 70 % of the board assembly costs [5]. An example cited was of an Ericsson cellphone resulting in a passive to active ratio of 21:1 [6]. The ratio has been reducing with the advancement in the integration technology. A complete RF front end for wireless local area network from Maxim Integrated Products, Inc. has passive device count of just 4 inductors, 33 capacitors, and 4 resistors [7].

Frequently used passive devices in several analog, RF, and mixed signal circuits include resistors, capacitors, inductors, varactors, etc. Resistors are used in voltage dividers, resistor arrays, biasing, etc. Capacitors are used in filters, tank circuits,

G. Haobijam and R. P. Palathinkal, *Design and Analysis of Spiral Inductors*,
DOI: 10.1007/978-81-322-1515-8_1, © Springer India 2014

to bypass or couple RF, as storage capacitor in DRAM, etc. Inductors are used in impedance matching, resonant circuits, filters, bias circuits, etc., of RF integrated circuits such as voltage-controlled oscillators (VCO), low noise amplifiers (LNA), mixers and power amplifiers. In order to reduce the size and realize low cost systems, today various passive devices are being integrated along with active devices on the same die. With the advancement of technology, various passive components are being offered that are integrated during the front end processing. Some of the types of resistors, capacitors, and inductors available in silicon technology are listed in Table 1.1 [1]. The values of the electrical parameters given here are as specified by the international technology roadmap for semiconductors (ITRS) in 2007 for on-chip passive devices in the RF and analog/mixed signal (AMS) chapter. These specifications are achievable with the currently available technology and tools. These clearly forecast the need of high quality passive devices. The available resistors are p-doped polysilicon resistors and back end of line (BEOL) metallization resistors. Polysilicon resistors are attractive due to a higher sheet resistance while BEOL resistors have less parasitics. Capacitors like metal oxide semiconductor (MOS), i.e., polysilicon-gated capacitors on single-crystal silicon, metal insulator metal (MIM), and metal oxide metal are offered in silicon technology. MIM is preferred because of its higher quality factor but it is less reliable as compared to MOS. Inductors offered are the planar and multilayer spiral inductors but quality factor of the inductor is generally low.

Passive devices are chosen depending on the specifications pertaining to the area of application and the technology adopted for implementation. The technology of choice for analog and mixed signal SoC is RF CMOS or SiGe-Bipolar CMOS. RF CMOS technology is preferred when the specification requirements are moderate and when there is a strong demand for cost reduction while SiGe-BiCMOS technology is preferred for specifications with higher sensitivity and low-power consumption requirements, with relatively low priority for cost reduction [8]. This will also depend on the time to market and overall system cost [2]. The integration of passive devices also requires an extra masking and processing steps. Therefore, the integration of passive devices in RF-SoC or mixed signal SoC plays a key role in determining the overall performance and processing cost and it offers various challenges [9].

1.2 On-Chip Inductor

Inductors are realized on-chip by laying out the metal trace on silicon using one or more metal interconnects in different ways. The most popular planar inductor topology is the square spiral. Figure 1.1 shows the spiral and its cross-section with the magnetic field lines. Since the metal turns are closely placed the flux in the turns and the flux lines passing through center of the coil are linked. The magnetic flux is defined as the magnetic field crossing the cross-sectional area of the conductor and is given by,

Table 1.1 Types of passive devices available in silicon technologies

Resistors	Sheet resistance ($\Omega/\square$)	σ Matching (%µm)	Temp. linearity (ppm per °C)	Parasitic capacitance (fF/µm^2)	
p^+ polysilicon	200–300	1.7	<100	0.1	
Thin film BEOL	50	0.2	<100	0.03	
Capacitors	Density (fF/µm^2)	Voltage linearity (ppm/V^2)	Leakage (A/cm^2)	σ Matching (%µm)	Q for 1pH, 5GHz
Metal oxide semiconductor	7	–	$<1e^{-9}$	–	–
Metal insulator metal	2	<100	$<1e^{-8}$	0.5	>50
Metal oxide metal	3.7	<100	–	–	–
Inductor	For a 1 nH inductor the achievable Q is around 29 at 5 GHz with a dedicated thick metal				

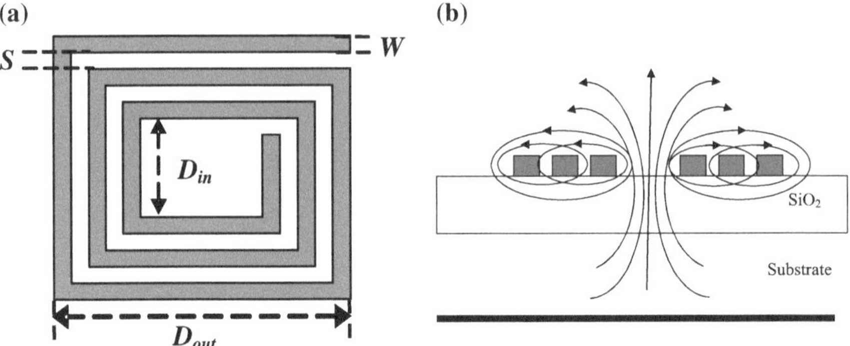

Fig. 1.1 Planar spiral inductor and its cross section showing the magnetic field lines

$$\begin{aligned}\Psi &= \oint B \cdot \mathrm{d}S \\ &= \mu \oint H \cdot \mathrm{d}S \end{aligned} \tag{1.1}$$

Inductance is defined as the ratio of the total flux linkages to the current to which they link. The self inductance is thus given by,

$$L = \frac{\Psi}{I} \tag{1.2}$$

where $L,\ \Psi,\ I,\ \mu,\ B,\ H,$ and $\mathrm{d}S$ are the inductance in henries (H), magnetic flux in webers (Wb), current in amperes (A), permeability in henries per meter (H/m), magnetic flux density expressed in tesla (Wb/m^2), magnetic field density in amperes per meter (A/m), and area in meters squared (m^2), respectively. Inductors will store energy from the applied voltage in their magnetic field through flux. The voltage induced is given by,

$$\begin{aligned}V &= \frac{\mathrm{d}\Psi}{\mathrm{d}t} \\ &= L\frac{\mathrm{d}i}{\mathrm{d}t}\end{aligned} \tag{1.3}$$

The mutual inductance caused by the magnetic interaction between two currents adds to the self-inductance. The mutual inductance on a turn i due to the impinging flux from the nearby turns j can be calculated as,

$$M_{ij} = \frac{\Psi_i}{I_j} \tag{1.4}$$

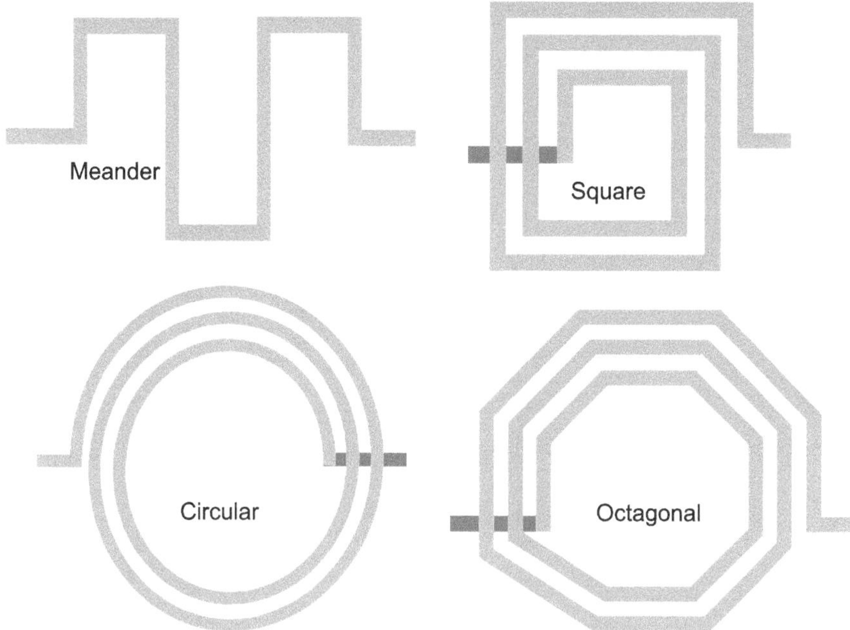

Fig. 1.2 Planar inductor structures

The total inductance is calculated as the sum of the self-inductance and mutual inductance. If the currents flows in the same direction the inductance increases, and if the current flows in the opposite direction the inductance decreases. The inductance computation for a spiral inductor is discussed in Chap. 2. If the turns of the spiral inductor are closely packed the mutual inductance due to the close coupling will be high. The turns of the spirals should be laid out so that the coupling is maximized. Other structures include meander, octagonal, circular, and solenoid as shown in Fig. 1.2. The geometrical or layout design parameters are the number of turns (N), spiral track width (W), track spacing (S), outer diameter (D_{out}), and inner diameter (D_{in}). The layout parameters are depicted in Fig. 1.2. The figure of merits (FOM) of on-chip spiral inductors are (i) quality factor, Q; (ii) optimum frequency, f_{max} at which Q reaches its maximum value, Q_{max}; (iii) self resonant frequency, f_{res} at which the inductor behaves like a parallel RC circuit in resonance [2], and (iv) inductance to silicon area ratio (L/A). A typical inductance and a quality factor plots are shown in Figs. 1.3 and 1.4, respectively. The inductance and the quality factor are frequency dependent. The calculation of the inductance and the quality factor is discussed later in Chap. 2. Qualitatively, three operating regions can be identified depending on the change of inductance values with frequency [10]. The regions are shown in the figure. Region I is the useful band of operation where the inductance value remains relatively constant. Region II is the transition region in which the inductance value changes at a faster rate and becomes negative. This frequency at

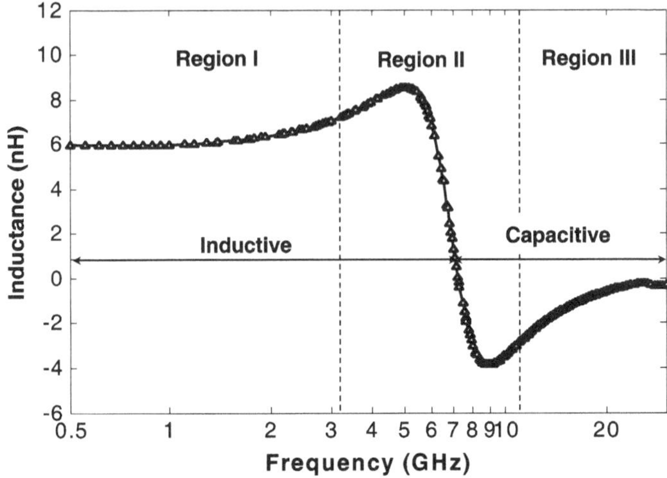

Fig. 1.3 Inductance as a function of frequency

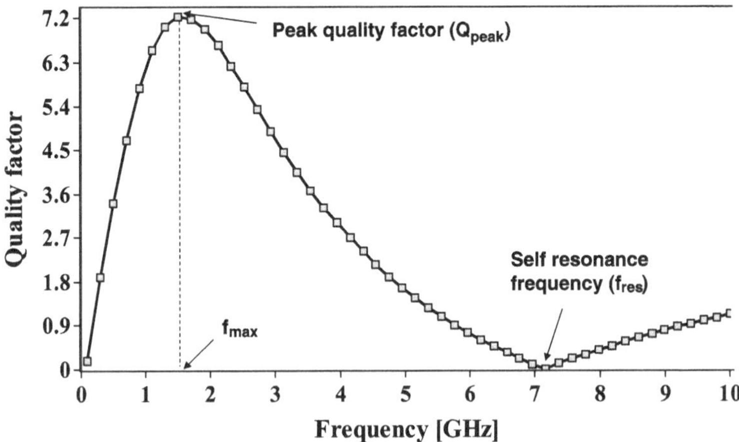

Fig. 1.4 Quality factor as a function of frequency

which the inductance value crosses zero is the first self-resonance frequency of the inductor. Beyond this frequency it enters Region III where the inductor resonates with its parasitic capacitance and is far from behaving as an inductor [2]. The inductor must not be used in this region.

The quality factor of integrated inductors on highly doped silicon substrate is quite low. This is mainly because of the loss in the conductor and the Si substrate. The losses in the conductor is proportional to the resistance of the metal. At low frequencies the series resistance will be given by

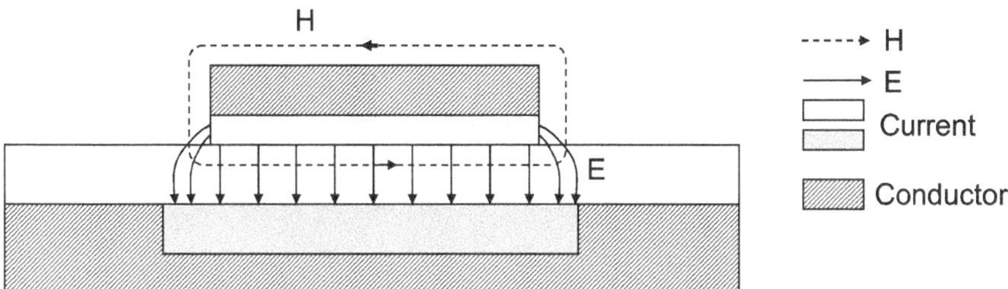

Fig. 1.5 Eddy current effect in the microstrip stripline conductor

$$R = \frac{\rho l}{W t} \tag{1.5}$$

where l is the total length of the metal, W is the width of the metal, t is the thickness of the metal, and ρ is the resistivity of the metal. At higher frequencies the series resistance becomes a complex function of frequency due to the skin effect. The high frequency current will recede to the bottom surface of the metal segment which is above the ground plane [11, 12]. This can be understood by considering the metal segments of the spiral inductor as microstrip transmission lines as shown in Fig. 1.5. As a result, the effective thickness of the metal decreases which is given by,

$$t_{\text{eff}} = \delta(1 - e^{\frac{-t}{\delta}}) \tag{1.6}$$

where δ is the skin depth of the metal. Therefore the equation of series resistance reduces to

$$R = \frac{l\rho}{W\delta(1 - e^{\frac{-t}{\delta}})} \tag{1.7}$$

So, resistance increases as the skin depth decreases with the frequency. At high frequency, the nonuniformity in the current will also result due to the magnetically induced eddy currents [13]. Consider a section of an n-turn circular inductor as shown in Fig. 1.6. Let the current in the inductor be I_{coil} and the associated magnetic flux be B_{coil}. The magnetic flux lines entering the page near the turn n will come out of the page in the center of the inductor. If the inner diameter of the spiral inductor is very small, this magnetic fields originating from current carrying outer metal turns will pass through the inner turns. According to Faraday's and Lenz's laws circular eddy currents will be induced in these inner metal turns of the inductor as shown in Fig. 1.6. An opposing magnetic field B_{eddy} will also be developed due to the eddy current. From the figure we can see that the eddy current will add to the I_{coil} on the inner side (left edge) and subtract from I_{coil} on the outer side (right edge) of the conductor. The current density will be thus, higher on the inner side than on the outer side and result in a nonuniform current in the metal turns of the spiral inductor.

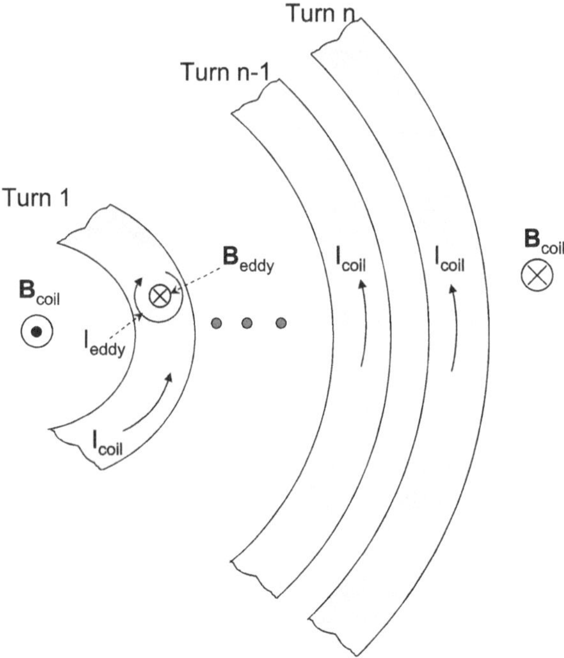

Fig. 1.6 Section of an n-turn circular inductor with the fields and the eddy current

The resistivity of silicon substrate ranges from 10 K Ω cm for lightly doped to 0.001 Ω cm for heavily doped substrate. Because of the low resistivity of the substrate, electric energy is coupled through the displacement current. This electrically induced displacement current flows vertically, perpendicular to the plane of the spiral inductor as shown in Fig. 1.7. Also, the magnetic field due to the inductor will penetrate through the substrate. This will induce eddy current loops that will oppose the excitation currents in the spiral turns and weaken the original magnetic field of the inductor.

With the scaling of CMOS technology, the number of metal layers and the total dielectric insulator thickness has increased. This has introduced new directions for performance improvements as the substrate coupling noise can be reduced with the increased distance of separation with scaling. Today, some RF CMOS technologies have thick top metal layer provision to reduce the series resistance and improve the quality factor of the inductor. However, the area occupied by inductors are quite large as compared to the area of active devices and it does not scale with the technology.

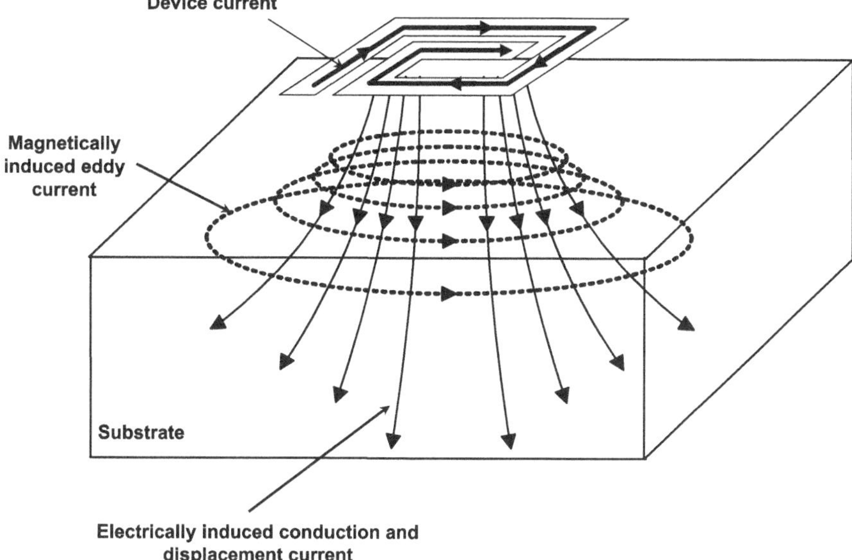

Fig. 1.7 Representation of the substrate currents in a spiral inductor. The *solid lines* represents the electrically induced currents and the *dashed lines* represent the eddy current [14]

1.3 Review of Si On-Chip Inductor Design and Optimization

The fabrication of inductors by integrated circuit techniques was investigated early in 1960s and 1970s but it was held that inductors are the most difficult component to integrate. This is because of the large chip area requirement for practical inductance values considered at several hundred megahertz and the low quality factor due to losses associated with the heavily doped silicon. Silicon on-chip inductor was first reported in 1990 by Nguyen and Meyer in a 0.8 μm silicon BiCMOS technology [15]. The inductors were square spirals of values 1.3 and 9.3 nH with a peak quality factor (Q) of 8 at 4.1 GHz and 3 at 0.9 GHz, respectively. They also proved its performance in an LC voltage controlled oscillator and RF bandpass filter circuits [16, 17]. Since 1990, there has been an enormous progress in the research on the performance trends, design and optimization, modeling, quality factor enhancement techniques, etc., of spiral inductors and significant results are reported in literature for various applications. Today, spiral inductors are widely used even at microwave frequencies and their applications in millimeter-wave circuits are investigated [4].

1.3.1 Spiral Inductor Structures

Most of the early efforts on the integration of inductor on Si seems to be especially inspired by the vision at that time to realize fully integrated CMOS radio transceivers. The first three papers by Nguyen and Meyer as mentioned above paved the way of research in this direction. In 1993, Chang et al. [18] showed that spiral inductors can be operated beyond the UHF band by reducing the capacitance and resistance loss with selective removal of the underlying substrate. A 100 nH square spiral inductor was designed with number of turns 20, metal width of 4, and 4 μm spacing, resulting in an outer dimension of 440 μm. Simulations on the SONNET EM 3D electromagnetic simulator showed that removal of the underlying substrate can increase the inductor self-resonance from 800 MHz to 3 GHz. The structure was fabricated through MOSIS as a standard n-well 2 μm CMOS IC. Data for the Q was not provided, but an equivalent circuit for the inductor at 800 MHz implies a Q of about 4 at that frequency. In 1994, Negus et al. [19] demonstrated an RF IC incorporating a monolithic inductor in a process that was claimed to be capable of producing inductors with Q's greater than 10. Also the integration of an inductor in a single-chip Global System for Mobile communications (GSM) transceiver RF integrated circuits were reported in [20, 21] but the measured data for the inductors are not given. For the first time the detail of inductor test and measurement were reported by Ashby et al. in 1994 [22]. Some 16 rectangular spiral inductors of metal trace widths 5, 9, 14, 19, 24, and 49 μm and different number of turns with same outer dimensions of 300 μm were fabricated in a high-speed complementary bipolar process and were characterized for use in wireless applications. The inductance value ranges from 1.74 to 35.4 nH with Q of 12 and 5.5, respectively. They also proposed a more accurate model modifying that of Nguyen and Meyer by adding extra components in the lumped equivalent model of the inductor.

In the following years active research on inductor integration continued leading to several innovative structures. These reports are grouped together according to the type of the inductor structure and presented here in the following subsections. They are not necessarily in the order of the year reported.

1.3.1.1 Spiral Inductor with Shunted Metal Layers

In standard silicon process, the thickness of the metal is limited to 1–2 μm and the series resistance loss is one of the major factors for low Q. The series resistance of the metal becomes a complex function at high frequencies due to eddy current effect and skin effect. In 1996, Soyuer et al. [23] proposed that the series resistance of the inductor could be minimized by increasing the thickness of the metal with shunting of multiple metal layers as in [24]. Different versions of four turn inductors, designed by shunting M2/M3, M3/M4, and M2/M3/M4 metal layers in parallel were reported. Thickness as high as 4 μm was achieved. All the three inductors have the same inductance around 2.2 nH but the inductor implemented by shunting M2/M3/M4 has the highest Q of 9.3 at 2.4 GHz. Similar results by the same authors are also given in [24]. It was also observed that Q did not increase in proportion with the metal thickness. This is because at higher frequencies above 2 GHz, the metal thickness may exceed skin depth and due to skin effect , the effective thickness of the metal will decrease. At 1 GHz the skin depth of Al, Cu, and Au are 2.56, 2.07, and 2.46 μm, respectively. The Q therefore did not increase in proportion to the thickness. In 2005, Chia-Hsin Wu et al. [25] presented another inductor where the turns of the inductor were shunted with selected metal layers. This configuration of the inductor structure demonstrated that the frequency at which the Q peaks can be modified by shunting metal layers selectively. This structure can be viewed as an inductor with increasing thickness from outer to innermost turns of the spiral. In fact, the series resistance will be reduced while the parasitic capacitance will be increased as the metal trace in the inner turns approach closer to the substrate. Therefore, Q at low frequency will be higher and due to larger parasitic capacitance Q decays early, resulting in a shift of the frequency at which Q peaks.

1.3.1.2 Multilevel Spiral Inductors

Planar structures require a minimum of two metal layers, with the spiral winding in one layer and the underpass in another metal layer. Planar spirals occupy a large area of the die. As the number of metal layers increases with the technology scaling, inductors can be realized exploiting multiple metal layers. In 1995, Merrill et al. [26] and Burghartz et al. [27] proposed multilevel inductors. Two or more spirals in different metal layers are connected in series as shown in Fig. 1.8 to increase the inductance to area ratio. It is commonly referred now as 'stacking' and the desired inductance can be realized in a smaller area. Further demonstrations followed in [10, 28–30]. Use of multiple metal layers has enabled different modifications in the inductor structure in order to increase the inductance to area ratio and realize inductors utilizing smaller area as compared to planar inductors or to enhance the performance for different applications.

Merrill et al. [26] observed that a three turn spiral inductor with M1/M2/M3 connected in series has nine times higher inductance to area ratio than the inductor with M1/M2/M3 connected in parallel. The series connected inductor was 16.7 nH while the parallel connected was only 1.84 nH. In [28] a spiral inductor built in M1/M2/M3/M4, with each spiral having 4 turns resulted in an inductance of 32 nH with a peak Q of 3 at 0.4 GHz and f_{res} of 1.8 GHz. The spirals in different layers may be placed directly one over the other so that they overlap exactly or slightly shifted diagonally to avoid overlapping. With maximum overlap, the inductance was higher since the spirals are coupled perfectly but the metal to metal overlap capacitance was higher, resulting in a smaller Q and f_{res}. In another study by Koutsoyannopoulos et al. [10], the layout parameters of the stacked spirals were varied to realize the same inductance with almost equal outer diameter. The diagonally shifted two layer spiral has an outer diameter of 281 μm, width and spacing of 9 μm and the exactly overlapping two layer spiral has an outer diameter of 286 μm, W of 14 μm and spacing of 9 μm. The inductors with exactly overlapping spirals have higher Q even though the capacitance between the two layers is higher. Since the metal width is larger, the series resistance was smaller. In 2001, Zolfaghari et al. [29] reported very high value inductance of 45 nH (M5,M4), 100 nH (M5,M4,M3), 180 nH (M5,M4,M3,M2,M1), and 266 nH (M5,M4,M3,M2) built in a 0.25 μm five metal layer process. They also showed that f_{res} can be increased by moving the spirals farther from each other. In a five metal layer process, two layer inductors with each spiral of seven turns, outer diameter of 240 m, W of 9 m, and spacing of 0.72 m were constructed in (M5, M4), (M5, M3) and (M5, M2). It was demonstrated that when the bottom spiral is moved from M4 to M2, the f_{res} increases from 0.96 to 1.79 GHz.

In 2002, Feng et al. [30] fabricated a super compact inductor in 0.18 μm process consisting of six identical spirals of four turns each, metal width of 1 μm, spacing of 0.5 μm and area of 22 μm × 23 μm. This inductor has an inductance of 10 nH and peak Q of 1.1 at 2.48 GHz. To further improve the f_{res}, miniature 3D inductor was proposed by Tang et al. [31]. In this structure, stacks of one turn spirals having different diameters are connected in series. It can be pictured as one stack placed inside the other. The metal to metal capacitance in this form of winding appears in series as opposed to parallel connection in stack and hence results in a smaller parasitic capacitance as compared to stack. The miniature 3D inductor increases f_{res} by 34 % with 8 % degradation in Q as compared to the stack of same inductance. Yin et al. analyzed this structure in detail in [32]. If the stack inductor has only one turn in each layer then it results in the vertical solenoid structure of Tsui and Lau [33] reported in 2005. It was shown that 4.8 nH vertical solenoid inductor approximately gives a 20 % increase in maximum Q and 50 % increase in f_{res}, while occupying only 20 % of the area as compared to 4.1 nH planar spiral inductor in a six-metal layer process. Earlier to this, a 5 nH horizontal solenoid inductor of 96 turns and area of 4 μm × 100 μm × 450 μm was reported by Edelstein and Burghartz [34]

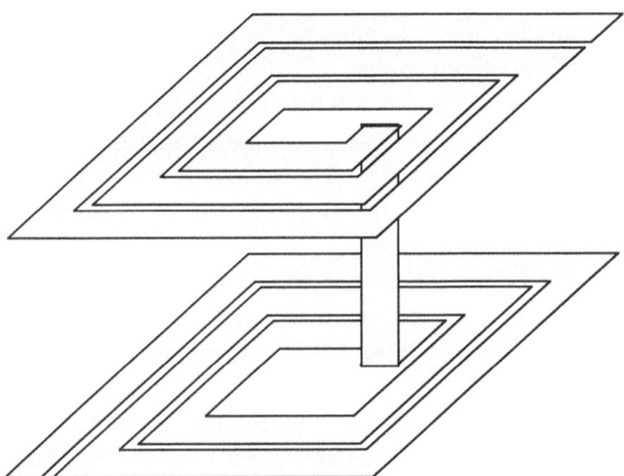

Fig. 1.8 Conventional two layer stack spiral inductor

in 1998 with a peak Q of 2.5 at 1.5 GHz. In summary, multilevel inductors have higher inductance to area ratio and occupy smaller area. Nevertheless, this reduced area is achieved at the cost of performance. The inter metal layer capacitance and the metal to substrate coupling capacitance increases and hence suffers from poor Q and smaller f_{res}.

1.3.1.3 Inductor with Patterned Ground Shield

In 1998, Yue and Wong [35] demonstrated that the silicon parasitics of on-chip inductor could also be eliminated with a patterned ground shield inserted between an on-chip spiral inductor and silicon substrate. The ground strips provide a good short to the electric field and terminate it before it reaches the silicon substrate. It was shown that at 1–2 GHz, the addition of a polysilicon patterned ground shield increases the inductor Q up to 33 % and reduces the substrate coupling between two adjacent inductors. However, the self-resonance frequency decreases due to the introduction of additional substrate parasitic capacitance. The effects of a ground shield shape and material on the performance of spiral inductors were studied in detail by Yim et al. [36]. They observed that with a PGS, the frequency dependence of the inductance increases while that of the series resistance decreases. The increase in the quality factor also depends on the area of the PGS, which means that there must be an optimum area of the PGS which gives the highest quality factor. They also compared the quality factor of inductors with n^+ buried/n-well layer PGS, metal-1 PGS and poly PGS. The inductor with poly PGS has the highest quality factor. Their investigation also showed that the isolation of adjacent inductors does not improve significantly with a PGS. Recently, Cheung et al. [37, 38] proposed a floating shield technique which has several advantages over the traditional ground shield. A differentially driven floating shield inductor has about 35 % improvement in Q-factor over an unshielded one.

1.3.1.4 Symmetric Inductors

In integrated circuits, the differential topology is preferred because of its less sensitivity to noise and interference. All the structures discussed above are asymmetric. For differential circuit implementation, a pair of planar spiral inductors can be used with their inner loops connected together in series [39] as shown in Fig. 1.9. Since the currents always flow in opposite directions in these two inductors, there must be enough spacing between them to minimize electromagnetic coupling. As a result, the overall area occupied by the inductors are very large. To eliminate the use of two inductors and reduce the chip area consumption, the center tapped spiral inductor was presented in 1995 by Kuhn et al. [40] for balanced circuits. Later, in 2002, Danesh and Long [41] presented a symmetrical inductor with enhanced Q for differential circuits as shown in Fig. 1.10. The symmetric inductor is realized by joining groups of coupled microstrip from one side of an axis of symmetry to the other using a

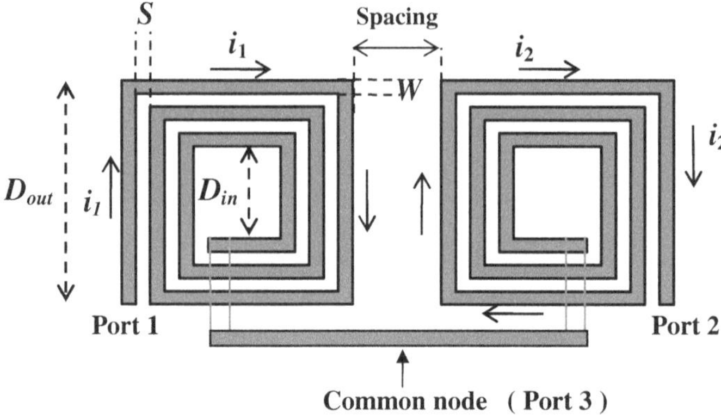

Fig. 1.9 Layout of a pair of asymmetric planar inductor for differential circuit implementation

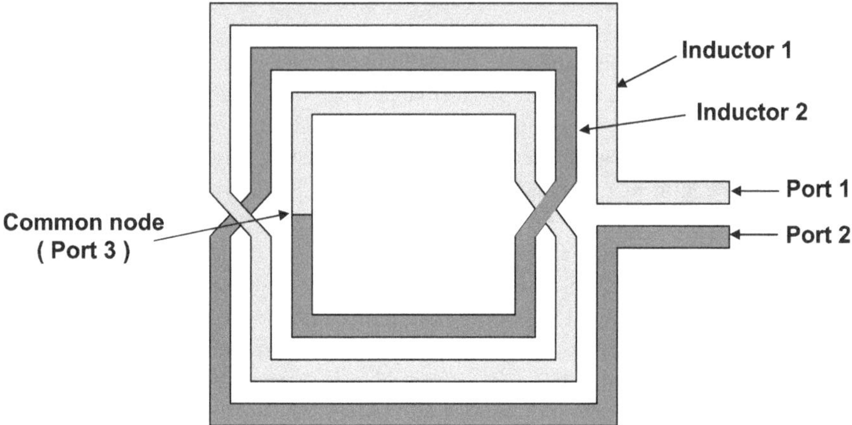

Fig. 1.10 Layout of a pair of asymmetric planar inductor for differential circuit implementation

number of cross-over and cross-under connections. The symmetrical inductor under differential excitation results in a higher Q and f_{res}. A 7.8 nH inductor with an outer diameter, metal width, and a spacing of 250, 8 and 2.8 μm, respectively has a peak Q of 9.3 at 2.5 GHz under differential excitation while Q is only 6.6 at 1.6 GHz under single ended condition. It also occupies less area than its equivalent asymmetrical pair of inductors. This type of winding of the metal trace was first applied to monolithic transformers by Rabjohn in 1991 [42]. A "group cross" symmetric inductor structure [43] manufactured on a printed circuit board (PCB), in which the metal traces cross each other in groups, was also shown to have less effective parasitic capacitances between two input ports and higher f_{res} and Q. However, the area of all these inductors is still large. Other different forms of symmetric windings were also studied [44–46].

1.3.1.5 Tapered or Variable Width Inductors

In 1997, Craninckx and Steyaert [39] studied that in a multiple turn planar inductor with a small radius, the largest contribution to the increase in the series resistance at high frequency comes from the inner turns. The magnetic field due the current in the inductor spiral passes through the inner turns which induces an electric field and thus generates the eddy current in the turns. Due to this eddy current, the current distribution in the inner conductor becomes nonuniform and hence increases the series resistance. They suggested that this can be prevented by making the width of the inner turns smaller than the outer ones or by using a 'hollow' coil, i.e., large radius. In 2000, Lopez-Villegas et al. [47] presented this approach in a more systematic manner by proposing a method to find the optimum width of each turn and achieve the maximum Q factor at a given frequency. This structure is generally referred now as tapered inductor. The improvement in the Q factor was reported for a micromachined 34 nH inductor reaching a Q of 17. By micromachining the substrate under the spiral inductor is removed. However, the importance of varying the width of the turns is not so significant for inductors on Si substrate where the substrate losses also come into effect at high frequency [13, 48].

1.3.2 Quality Factor Enhancement Techniques

On-chip spiral inductors fabricated on Si substrates suffer from poor quality factor due to ohmic and substrate losses. The quality factor is inversely proportional to the finite resistance of the metal layer which becomes a complex function at high frequencies and the losses in inductors increase as a result of induced currents and dielectric losses. The low resistivity of silicon substrate results in capacitive coupling, allowing the flow of conduction current through the substrate. Several techniques have been used to enhance the quality factor of inductors on silicon. One such method is micromachining, i.e., etching out the silicon underneath the inductor using front side etching or backside etching or by using high aspect ratio and surface micromachining techniques as reported in [18, 49–55], etc. These methods result in near elimination of the substrate loss thereby yielding very high quality factor inductors with high self resonant frequency. In 1998, Yue and Wong [35] demonstrated that the silicon parasitics of on-chip inductor could also be eliminated with a patterned ground shield as discussed above in Sect. 1.3.1.3. Other methods include the use of high resistivity silicon substrates [56] and sapphire substrates [57], differential excitation technique [41] as discussed in Sect. 1.3.1.4, multilayer substrate [58], n-well formation [59], the formation of porous silicon [60], proton bombardment [61], etc. Most of these processes are uncommon in digital logic CMOS process.

1.3.3 Inductor Design and Optimization Methods

The performance of a spiral inductor is determined by its geometrical or layout parameters and the technological parameters. The dependence of the quality factor and inductance on these parameters have been studied in detail [10, 62, 63]. Thus, the complexity in the design of an on-chip inductor lies in deciding these layout parameters in order to achieve the target inductance with its desired quality factor. Various methods have been proposed to design and optimize an inductor. In 1998, Niknejad and Meyer [64] developed a computer-aided design tool 'ASITIC' (Analysis and simulation of spiral inductors and transformers for ICs) for designing, optimizing, and modeling of the spiral inductor and transformers. It allows the user to search the parameters space of an inductor optimization problem while trading off between speed and accuracy. It gives a SPICE file which can be used in circuit analysis and the layout of the spiral inductor can also be exported. In 1999, Hershenson et al. [65] presented an efficient optimization methodology based on the 'geometric programming (GP)' [66]. The authors showed that the inductor design goal, i.e., to optimize the Q factor can be formulated as a geometric program to obtain the trade off curve between L and Q for a particular operating frequency. Such a curve aids the designer in deciding the inductance value and explore the trade offs of performance for a particular application. In 2000, Post [67] developed an algorithm to find the optimized layout parameter based on the well accepted model of Yue and Wong [35]. Similarly, other optimization methods were proposed based on sequential quadratic programming [68, 69], simulated annealing [70], artificial neural network [71, 72], etc., which have proved to be more efficient reducing the computation time and converging rapidly to the optimal design.

From the previous sections of this chapter, we have seen that with the advancement of the Si technology various inductor structures have evolved from asymmetric to symmetric and from planar to multilayer to meet the demands of high performance miniaturized circuits. The ohmic loss and the substrate loss can be minimized in various ways. The design is a complex process involving the optimization of its layout parameters, using various tools and methodologies available today to cater to the needs of the design and reduce the design time to market cycle.

1.4 Summary

In this chapter, the design of on-chip inductor was discussed with a review on its innovative structure evolution and design trends, followed by a discussion of the unsolved problems and scope of this work. A brief summary on the trend of integrated passive devices was given. The first and basic step in the design of integrated inductor is the selection of a particular topology and layout in a chosen process technology. With the scaling of CMOS technology, inductor structures of various shapes have evolved from asymmetric to symmetric and from planar to multilayer.

A review of different inductor structures fabricated in standard CMOS process was given. The impact of different ways of winding on performance metrics like quality factor, inductance, and self resonance frequency was discussed. Several techniques reported in order to improve the quality factor by reducing the substrate losses were also reviewed. One requires a good understanding of the performance trends with respect to the layout and process parameters to optimize the design. A review of such optimization methodologies was also included. Several issues on the design trends and optimization methodologies that have motivated us for this work are discussed.

References

1. RF and AMS technologies for wireless communications. The International Technology Roadmap For Semiconductors, 2007 edn. http://www.itrs.net/Links/2007ITRS/Home2007.htm
2. Bennett, H.S., Brederlow, R., Costa, J.C., Cottrell, P.E., Huang, W.M., Immorlica, A.A., Mueller, J.E., Racanelli, M., Shichijo, H., Weitzel, C.E., Zhao, B.: Device and technology evolution for silicon-based RF integrated circuits. IEEE Trans. Electron Dev. **52**(7), 1235–1258 (2005)
3. Abidi, A.A.: Rf cmos comes of age. IEEE J. Solid-State Circuits **39**(4), 549–561 (2004)
4. Dickson, T., LaCroix, M.-A., Boret, S., Gloria, D., Beerkens, R., Voinigescu, S.: 30–100–GHz inductors and transformers for millimeter-wave (Bi)CMOS integrated circuits. IEEE Trans. Microw. Theory Tech. **53**(1), 123–133 (2005)
5. Pedder, D.J.: Technology and infrastructure for embedded passive components. On Board Technology, pp. 8–11 (2005)
6. Ulrich, R.K., Brown, W.D., Ang, S.S., Barlow, F.D., Elshabini, A., Lenihan, T.G., Naseem, H.A., Nelms, D.M., Parkerson, J., Schaper, L.W., Morcan, G.: Getting aggressive with passive devices. IEEE Circuits Devices Mag. **16**(5), 16–25 (2000)
7. 11b WLAN transceiver shrinks circuit board and bill of material. Application note. http://www.eetindia.co.in
8. Matsuzawa, A.: RF-Soc—Expectations and required conditions. IEEE Trans. Microw. Theory Tech. **50**(1), 245–253 (2002)
9. Dunn, J.S., Freeman, G., Greenberg, D.R., Groves, R.A., Guarin, F.J., Hammad, Y., Joseph, A.J., Lanzerotti, L.D., Onge, S.A.S., Orner, B.A., Rieh, J.-S., Stein, K.J., Voldman, S.H., Wang, P.-C., Zierak, M.J., Subbanna, S., Harame, D.L., Herman, D.A., Meyerson, J.B.S.: Foundation of RF CMOS and SiGe BiCMOS technologies. IBM J. Res. Dev. **47**(2/3), 101–138 (2003)
10. Koutsoyannopoulos, Y.K., Papananos, Y.: Systematic analysis and modeling of integrated inductors and transformers in RFIC design. IEEE Trans. Circuits Syst. II **47**(8), 699–713 (2000)
11. Pucel, R., Massè, D., Hartwig, C.: Losses in microstrip. IEEE Trans. Microw. Theory Tech. **16**(6), 342–350 (1968)
12. Faraji-Dana, R., Chow, Y.: The current distribution and ac resistance of a microstrip structure, IEEE Trans. Microw. Theory Tech. **38**(9), 1268–1277 (1990)
13. Bahl, I.J.: Lumped Elements for RF and Microwave Circuits. Artech House Publishers, Norwood (2003)
14. Niknejad, A.M., Meyer, R.G.: Design, Simulation and Applications of Inductors and Transformers for Si RF ICs. Kluwer Academic Publishers, Boston (2000)
15. Nguyen, N.M., Meyer, R.G.: Si IC-compatible inductors and LC passive filters. IEEE J. Solid-State Circuits **25**(4), 1028–1031 (1990)

16. Nguyen, N.M., Meyer, R.G.: A Si bipolar monolithic RF bandpass amplifier. IEEE J. Solid State Circuits. **27**(1), 123–127 (1992)
17. Nguyen, N.M., Meyer, R.G.: A 1.8-GHz monolithic LC voltage-controlled oscillator. IEEE J. Solid State Circuits. **27**(3) 444–450 (1992)
18. Chang, J.Y.C., Abidi, A.A., Gaitan, M.: Large suspended inductors on silicon and their use in a 2μm CMOS RF amplifier. IEEE Electron Dev. Lett. **14**(5), 246–248 (1993)
19. Negus, K., Koupal, B., Wholey, J., Carter, K., Millicker, D., Snapp, C., Marion, N.: Highly-integrated transmitter rfic with monolithic narrowband tuning for digital cellular handsets. IEEE Int. Solid State Circuits Conf. Dig. Tech. Pap. 38–39 (1994)
20. Stetzler, T.D., Post, I.G., Havens, J.H., Koyama, M.: A 2.7-4.5 V single chip GSM transceiver RF integrated circuit. IEEE J. Solid-State Circuits **30**(12), 1421–1429 (1995)
21. Marshall, C., Behbahani, F., Birth, W., Fotowai, A., Fuchs, T., Gaethke, R., Heimeri, E., Lee, S., Moore, P., Navid, S., Saur, E.: A 2.7 v gsm transceiver ics with on-chip filtering. IEEE Int. Solid State Circuits Conf. Dig. Tech. Pap. 148–149 (1995)
22. Ashby, K.B., Finley, W.C., Bastek, J.J., Moinian, S., Koullias, I.A.: High Q inductors for wireless applications in a complementary silicon bipolar process. In: Proceedings of the Bipolar/BiCMOS Circuits and Technology Meeting, Oct 1994
23. Soyuer, M., Burghartz, J.N., Jenkins, K.A., Ponnapalli, S., Ewen, J.F., Pence, W.E.: Multi-level monolithic inductors in silicon technology. Electron. Lett. **31**(5), 359–360 (1995)
24. Burghartz, J.N., Soyuer, M., Jenkins, K.A.: Microwave inductors and capacitors in standard multilevel interconnect silicon technology. IEEE Trans. Microw. Theory Tech. **44**(1), 100–104 (1996)
25. Wu, C.-H., Kuo, C.-Y., Liu, S.-I.: Selective metal parallel shunting inductor and its VCO application. IEEE Trans. on Circuits Syst. I: Regul. Pap. **52**(9) 1811–1818 (2005)
26. Merrill, R.B., Lee, T.W., You, H., Rasmussen, R., Moberly, L.A., Optimization of high Q integrated inductors for multilevel metal CMOS. IEDM Tech. Dig. 983–986 (1995)
27. Burghartz, J.N., Soyuer, M., Jenkins, K.A., Hulvey, M.D.: High-Q inductors in standard silicon interconnect technology and its application to an integrated RF power amplifier. IEDM Tech. Dig. 1015–1018 (1995)
28. Burghartz, J.N., Jenkins, K.A., Soyuer, M.: Multilevel-spiral inductors using VLSI interconnect technology. IEEE Electron Dev. Lett. **17**(9) 428–430 (1996)
29. Zolfaghari, A., Chan, A., Razavi, B.: Stacked inductors and transformers in CMOS technology. IEEE J. Solid-State Circuits **36**(4), 620–628 (2001)
30. Feng, H., Jelodin, G., Gong, K., Zhan, R., Wu, Q., Chen, C., Wang, A.: Super compact RFIC inductors in 0.18 μm CMOS with copper interconnects. IEEE MTT-S Int. Microwave Symp. Dig. **1**, 553–556 (2002)
31. Tang, C.-C., Wu, C.-H., Liu, S.-I.: Miniature 3-D inductors in standard CMOS process. IEEE J. Solid-State Circuits **37**(4), 471–480 (2002)
32. Yin, W.-Y., Xie, J.-Y., Kang, K., Shi, J., Mao, J.-F., Sun, X.-W.: Vertical topologies of miniature multispiral stacked inductors. IEEE Trans. Microw. Theory Tech. **56**(2), 475–486 (2008)
33. Tsui, H.-Y., Lau, J.: An on-chip vertical solenoid inductor design for multigigahertz CMOS RFIC. IEEE Trans. Microw. Theory Tech. **53**(6), 1883–1890 (2005)
34. Edelstein, D.C., Burghartz, J.N.: Spiral and solenoidal inductor structures on silicon using Cu-damascene interconnects. In: Proceedings of the IEEE International Interconnect Technology Conference, June 1998
35. Yue, C.P., Wong, S.S.: On-chip spiral inductors with patterned ground shields for Si-based RF IC's. IEEE J. Solid-State Circuits **33**(5), 743–752 (1998)
36. Yim, S.-M., Chen, T., Kenneth, K.O.: The effects of a ground shield on the characteristics and performance of spiral inductors. IEEE J. Solid-State Circuits **37**(2), 237–244 (2002)
37. Cheung, T.S.D., Long, J.R., Vaed, K., Volant, R., Chinthakindi, A., Schnabel, C.M., Florkey, J., Stein, K.: Differentially shielded monolithic inductors. In: Proceedings of the IEEE Custom Integrated Circuits Conference, Sept 2003
38. Cheung, T.S.D., Long, J.R.: Shielded passive devices for silicon-based monolithic microwave and millimeter-wave integrated circuits. IEEE J. Solid-State Circuits **41**(5), 1183–1200 (2006)

39. Craninckx, J., Steyaert, M.: A 1.8-GHz low-phase-noise CMOS VCO using optimized hollow spiral inductors. IEEE J. Solid-State Circuits **32**(5), 736–744 (1997)
40. Kuhn, W.B., Elshabini-Riad, A., Stephenson, F.W.: Centre-tapped spiral inductors for monolithic bandpass filters. Electron. Lett. **31**(8), 625–626 (1995)
41. Danesh, M., Long, J.R.: Differentially driven symmetric microstrip inductors. IEEE Trans. Microw. Theory Tech. **50**(1), 332–341 (2002)
42. Rabjohn, G.G.: Monolithic microwave transformers. Master's thesis. Carleton Univ, Ottawa, ON, Canada (1991)
43. Wang, Y.-Y., Li, Z.-F.: Group-cross symmetrical inductor (GCSI): a new inductor structure with higher self-resonance frequency and Q factor. IEEE Trans. Magn. **42**(6), 1681–1686 (2006)
44. Chen, W.-Z., Chen, W.-H.: Symmetric 3D passive components for RF ICs application. In: Proceedings of the IEEE Radio Frequency Integrated Circuits (RFIC), Symposium, pp. 599–602, June 2003
45. Kodali, S., Allstot, D.J.: A symmetric miniature 3D inductor. In: Proceedings of the International Symposium on Circuits and Systems, vol. 1, pp. I–92, May 2003
46. Yang, H.Y.D.: Design considerations of differential inductors in CMOS technology for RFIC. In: Proceedings of the IEEE Radio Frequency Integrated Circuits (RFIC), Symposium, pp. 449–452, June 2004
47. Lopez-Villegas, J.M., Samitier, J., Cane, C., Losantos, P., Bausells, J.: Improvement of the quality factor of RF integrated inductors by layout optimization. IEEE Trans. Microw. Theory Tech. **48**(1), 76–83 (2000)
48. Niknejad, A.M., Meyer, R.G.: Design, simulation and applications of inductors and transformers for SI RF ICs. Kluwer Academic, Boston (2000)
49. Chi, C.-Y., Rebiez, G.M.: Planar microwave and millimeter-wave lumped elements andcoupled-line filters using micro-machining techniques. IEEE Trans. Microw. Theory Tech. **43**(4), 730–738 (1995)
50. Ozgur, M., Zaghloul, M.E., Gaitan, M.: High Q backside micromachined CMOS inductors. In: Proceedings of the IEEE International Symposium on Circuits and Systems, vol 2, pp. 577–580, May 1999
51. Hongrui, J., Wang, Y., Yeh, J.-L.A., Tien, N.C.: On-chip spiral inductors suspended over deep copper-lined cavities. IEEE Trans. Microw. Theory Tech. **48**(12) 2415–2423 (2000)
52. Lakdawala, H., Zhu, X., Luo, H., Santhanam, S., Carley, L.R., Fedder, G.K.: Micromachined high-Q inductors in a 0.18μm copper interconnect low-k dielectric CMOS process. IEEE J. Solid State Circuits **37**(3) 394–403 (2002)
53. Hsieh, M.-C., Fang, Y.-K., Chen, C.-H., Chen, S.-M., Yeh, W.-K.: Design and fabrication of deep submicron CMOS technology compatible suspended high-Q spiral inductors. IEEE Trans. Electron Dev. **51**(3), 324–331 (2004)
54. Ribas, R.P., Lescot, J., Leclercq, J.-L., Karam, J.M., Ndagijimana, F.: Micromachined microwave planar spiral inductors and transformers. IEEE Trans. Microw. Theory Tech. 48(8) 1326–1335 (2000)
55. Park, J.Y., Allen, M.G.: High Q spiral-type microinductors on silicon substrates. IEEE Trans. Magnetics **35**(5) 3544–3546 (1999)
56. Ashby, K.B., Koullias, I.A., Finley, W.C., Bastek, J.J., Moinian, S.: High-Q inductors for wireless applications in a complementary silicon bipolar process. IEEE J. Solid State Circuits **31**(1), 4–9 (1996)
57. Burghartz, J.N., Edelstein, D.C., Jenkins, K.A., Kwark, Y.H., Spiral inductors and transmission lines in silicon technology using copper-damascene interconnects and low-loss substrates. IEEE Trans. Microw. Theory Tech. **45**(10, Part 2) 1961–1968 (1997)
58. Okabe, H., Yamada, H., Yamasaki, M., Kagaya, O., Sekine, K., Yamashita, K.: Characterization of a planar spiral inductor on a composite-resin low-impedance substrate and its application to microwave circuits. IEEE Trans. Compon. Packag. Manuf. Technol. Part B **21**(3) 269–273 (1998)
59. Kihong K., Kenneth, O.: Characteristics of an integrated spiral inductor with an underlying n-well. IEEE Trans. Electron Dev. **44**(9) 1565–1567 (1997)

60. Choong-Mo, N., Young-Se, K.: High-performance planar inductor on thick oxidized porous silicon (OPS) substrate. IEEE Microw. Guided Wave Lett. **7**(8) 236–238 (1997)
61. Lee, L.S., Chungpin, L., Liao, C., Lee, C.-L., Huang, T.-H., Tang, D.D.-L., Duh, T.S., Yang, T.-T.: Isolation on si wafers by MeV proton bombardment for RF integrated circuits. IEEE Trans. Electron Dev. **48**(5), 928–934 (2001)
62. Long, J.R., Copeland, M.A.: The modeling, characterization, and design of monolithic inductors for silicon RF IC's. IEEE J. Solid State Circuits **32**(3), 357–369 (1997)
63. Andersen, R.B., Jorgensen, T., Laursen, S., Kolding, T.E.: EM-simulation of planar inductor performance for epitaxial silicon processes. Analog Integr. Circ. Sig. Process **30**(1), 51–58 (2002)
64. Niknejad, A.M., Meyer, R.G.: Analysis, design, and optimization of spiral inductors and transformers for Si RF IC's. IEEE J. Solid State Circuits **33**(10), 1470–1481 (1998)
65. Hershenson, M.M., Mohan, S.S., Boyd, S.P., Lee, T.H.: Optimization of inductor circuits via geometric programming. In: Proceedings of the Design Automation Conference, June 1999
66. Duffin, R.J., Peterson, E.L., Zener, C.: Geometric programming-theory and application. Wiley, New York (1967)
67. Post, J.E.: Optimizing the design of spiral inductors on silicon. IEEE Trans. Circuits and Syst. II: Analog and Digital Signal Process. **47**(1), 15–17 (2000)
68. Zhan, Y., Sapatnekar, S.S.: Optimization of integrated spiral inductors using sequential quadratic programming. In: Proceedings of the IEEE Design, Automation and Test in Europe Conference and Exhibition, vol 1, pp. 622–627, Feb 2004
69. Nieuwoudt A., Massoud, Y., Variability-aware multilevel integrated spiral inductor synthesis. IEEE Trans. Comput. Aided Design Integr. Circuits Syst. **25**(12) 2613–2625 (2006)
70. Bhaduri, A., Vijay, V., Agarwal, A., Vemuri, R., Mukherjee, B., Wang, P., Pacelli, A.: Parasitic-aware synthesis of RF LNA circuits considering quasi-static extraction of inductors and interconnects. In: Proceedings of the 47th Midwest Symposium on Circuits and System, vol 1, pp. 477–480, July 2004
71. Mukherjee, S., Mutnury, B., Dalmia, S., Swaminathan, M.: Layoutlevel synthesis of RF inductors and filters in LCP substrates for Wi-Fi applications. IEEE Trans. Microw. Theory Tech. **53**, 2196–2210 (2005)
72. Mandal, S.K., Sural, S., Patra, A.: ANN- and PSO-based synthesis of on-chip spiral inductors for RF ICs. IEEE Trans. Comput. Aided Design Integr. Circuits Syst. **27**(1) 188–192 (2008)

Chapter 2
Optimization of Spiral Inductor with Bounding of Layout Parameters

2.1 Introduction

A typical spiral inductor design problem is to determine its optimum layout parameters for a given inductance that will result in the highest quality factor at desired frequency. This chapter discusses new a approach for spiral inductor design and its optimization. Section 2.2 proposes an algorithm to decide the bounds on the design parameters of spiral inductor for a large range of physical inductance values that satisfies a given area specification. With this parameter bounds, we can eliminate a large proportion of the redundant sample designs. Section 2.3 presents an extensive analysis of the dependence of quality factor, peak frequency, self resonance frequency, and area of a spiral inductor on its layout parameters, while keeping the inductance value constant as opposed to various studies reported. The benefits of such a study is discussed and illustrated with a design example. In Sect. 2.4, it is proved that by incorporating this bounding technique the feasible region of the problem can be determined and the number of function evaluations required to converge to the optimum solution can be reduced. Hence optimization can be completed in a few seconds efficiently. Numerical algorithms based on lumped element model are generally adopted, since EM simulations are computationally expensive and time-consuming. However, EM simulators provide the most accurate design. An optimization using an EM simulator would be acceptable only if few structures have to be simulated to find the optimum design parameters. Such an algorithm is proposed in Sect. 2.5, which consists of the minimum steps required to design and optimize a spiral inductor by simulating a few inductor structures using a 3D EM simulator for a given technology. The selection of the few structures is based on the insights obtained from the studies of performance trends of the previous section. The chapter is summarized in Sect. 2.6.

G. Haobijam and R. P. Palathinkal, *Design and Analysis of Spiral Inductors*,
DOI: 10.1007/978-81-322-1515-8_2,

2.2 Bounding of Layout Parameters

A spiral inductor optimization problem may be formulated as

$$\begin{aligned} \text{maximize} \quad & Q(N, W, D, S) \\ \text{subject to} \quad & L(N, W, D, S) \leq L_{desired} \\ & N_{min} \leq N \leq N_{max} \\ & W_{min} \leq W \leq W_{max} \\ & S_{min} \leq S \leq S_{max} \\ & D_{min} \leq D \leq D_{max} \end{aligned}$$

where $Q(N, W, D, S)$ is the objective function and N, W, D and S are the optimization variables. The set of sample points for which the objective function and all constraints are defined is the domain of the optimization problem and the set of all points that satisfies all the constraints is the feasible set. The size of the design search space and the number of function evaluation are determined by the lower and upper bounds on these variables. For fewer function evaluation it is important to restrict the search space only to the feasible region. This means that only the range of N, W, D, and S, which will result in the desired value of inductance, must be specified to the optimizer. In this section a method of bounding on these optimization variables is demonstrated and locate the feasible region for any desired value of inductance.

The spiral inductor design variables N, W, D and S are not independent. The limits on the outer diameter will decide the possible combinations of N, W and S governed by the relation

$$D_{out} = D_{in} + 2\,W\,N + 2\,S\,(N - 1) \tag{2.1}$$

Therefore to simplify, it is assumed that the spiral inductor outer diameter is specified. For any desired inductance value several combinations of the N, W, D_{out} or D_{in} and S exist. Also, there will certainly be a range of inductance values that satisfy the same area limitation. The algorithm develops the spiral inductor layout parameter bound curves of all such inductors and these curves can then be used to determine the bound on the number of turns and width for any value of inductance that can be designed satisfying the same area limitation [1]. The algorithm is explained by the flowchart in Fig. 2.1 and it consists of three major steps as given below:

(i) Determine the maximum number of turns, N_{max} that can be accommodated in the limited area for each width and spacing of the spiral.

(ii) Keep the outer diameter, D_{out} at maximum and constant. For each width, W and spacing, S, vary the number of turns from 1 to N_{max}, keeping $D_{in} \geq D_{in\ min}$ and compute the inductance for each case. One may consider that the turns of the inductor are spiraling in gradually. Therefore, in each combination D_{in} will vary and will be at its maximum limit for each N, W and S combination.

(iii) Keep the inner diameter, D_{in} minimum and constant. For each width and spacing, vary the number of turns from 1 to N_{max}, keeping $D_{out} \leq D_{out\ max}$ and compute the inductance for each case again. Here, we may consider that the

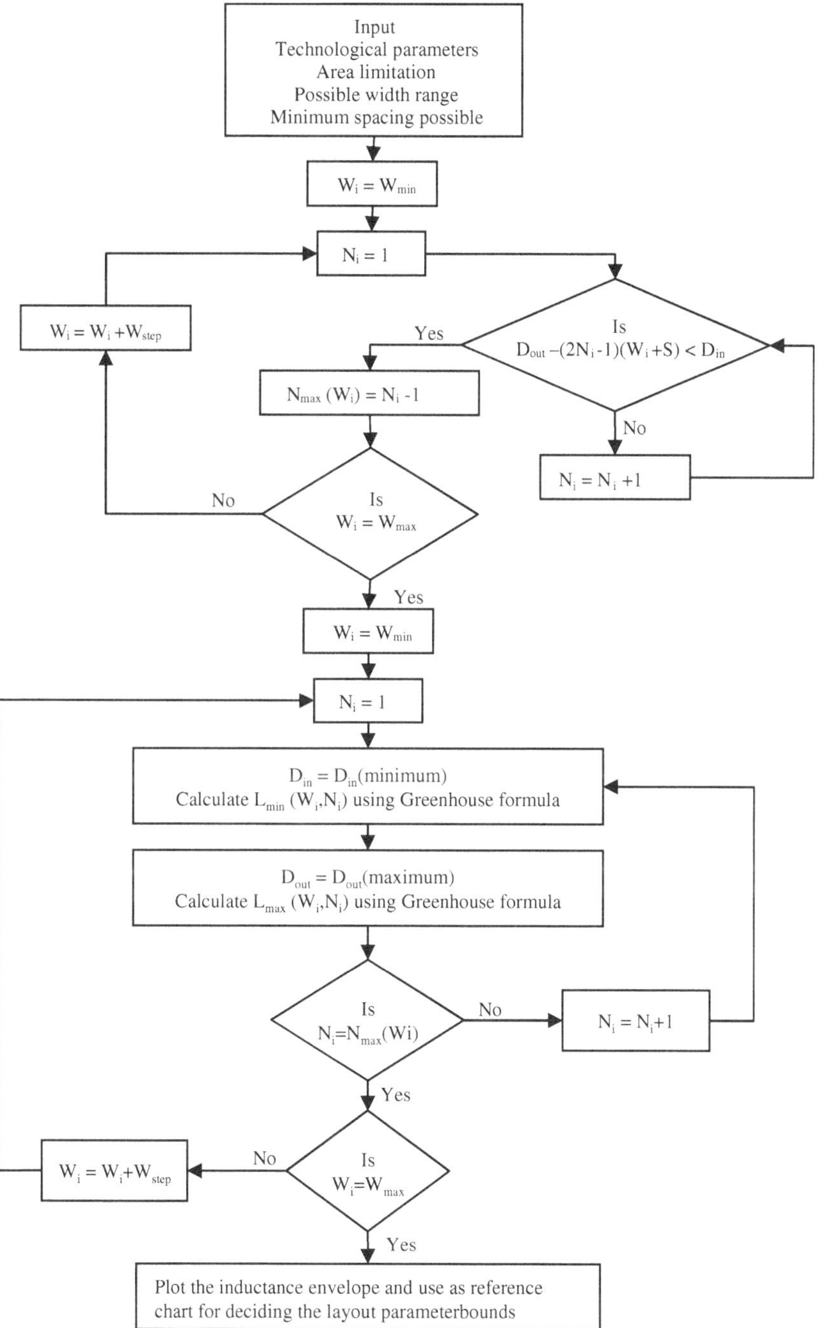

Fig. 2.1 Flowchart to determine the layout parameter bounds of spiral inductor

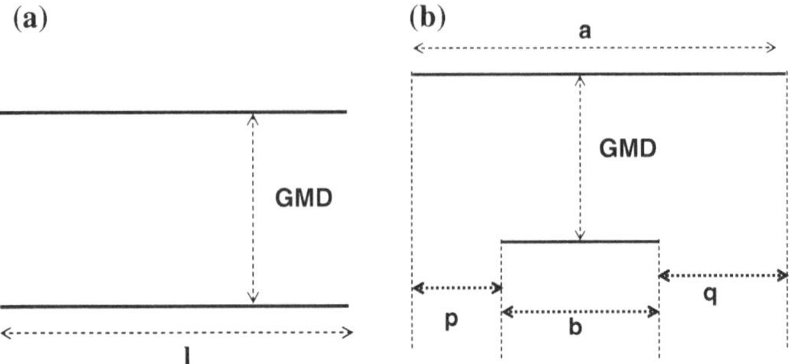

Fig. 2.2 Parallel conductors of **a** equal length **b** different length

turns of the inductor are spiraling out gradually. In this case, D_{out} will vary for each N, W and S combination within the area limit.

The inner or outer diameter is given by Eq. 2.1. In this way, for each N, W and S combination we will get the maximum inductance from step (ii) and minimum inductance from step (iii) by varying D_{in} and D_{out} within the area limits. Different formulae are proposed in the literature to compute the inductance of a spiral inductor. In this work we followed the inductance calculation algorithm developed by Greenhouse [2] based on Grover's [3] formula. In this method the planar spiral is divided into a number of straight conductor segments. The total inductance is calculated as the sum of all the self-inductance of the straight segments and mutual inductance, both positive and negative between the parallel segments. For example, a square inductor of three turns can be divided into 12 segments. The number of segments may not necessarily be a multiple of four. So the inductance is calculated as

$$L_{total} = L_{self} + M_{+} + M_{-} \tag{2.2}$$

where L_{total} is the total inductance, L_{self} is the total sum of self-inductance of all the segments, M_{+} and M_{-} are the sum total of all the positive and negative mutual inductances of all the segments. The self-inductance of a segment is calculated as

$$L = 0.002\, l \left[\ln \frac{2l}{W+t} + 0.50049 + \frac{W+t}{3\, l} \right] \tag{2.3}$$

where l is the length of the segment, W is the width of the conductor, and t is the thickness of the conductor. The mutual inductance between two parallel conductors of equal length as shown in Fig. 2.2a is given by

$$M = 2lH \tag{2.4}$$

where l is the length of the conductor and H is the mutual inductance parameter given by

$$H = \ln\left\{\frac{l}{GMD} + \sqrt{1 + \frac{l^2}{GMD^2}}\right\} - \sqrt{1 + \frac{GMD^2}{l^2}} + \frac{GMD}{l} \tag{2.5}$$

where GMD is the geometric mean distance between the two conductors. This is approximately equal to the distance, d between the center of the conductors. Its exact value is calculated as

$$\ln GMD = \ln d - \left\{\frac{W^2}{12\, d^2} + \frac{W^4}{60\, d^4} + \frac{W^6}{168\, d^6} + \frac{W^8}{360\, d^8} + \cdots\right\} \tag{2.6}$$

For the spiral inductor case, the length of the parallel conductors is not equal such as the case shown in Fig. 2.2b. If a and b are the length of the two conductors separated by GMD as shown in the figure, their mutual inductance is calculated as

$$2\, M_{a,b} = \{M_{b+p} + M_{b+q}\} - \{M_p + M_q\} \tag{2.7}$$

where each mutual terms are calculated using Eq. 2.4. For example,

$$M_{b+p} = 2l_{b+p}H_{b+p} = 2(b + p)H_{b+p} \tag{2.8}$$

where H_{b+p} is the mutual inductance parameter given by Eq. 2.5 for $l = b + p$. Thus the inductance is calculated summing all the self and the positive and negative mutual inductances of all the conductor segments of the spiral inductor.

To illustrate the bounding methodology we consider here an example, where D_{out} is assumed to be 400 μm. The width was chosen to vary from 5 to 25 μm. Several studies [4, 5], have shown that the tight coupling of the magnetic field maximizes the quality factor and reduces the chip area for a given inductor layout. The interwinding capacitance from tighter coupling has only a slight impact on performance. Therefore, the spacing was kept constant at 2 μm. The largest $N_{\max}$ was found to be 26. The possible inductances vary from 0.13 to 140 nH. The minimum and maximum inductances of all possible combinations of N, W and S is shown in Fig. 2.3a and b, respectively. In the figure, inductance values only for number of turns up to 10 are shown and D_{in} was allowed to be as small as 50 μm. The information from Fig. 2.3a and b is combined to generate the layout parameter bound curves as shown in Fig. 2.4. The curves are plotted only for widths 5, 10, 15, 20, and 25 μm for clarity. The other widths that are not shown in the figure also follow the same pattern. In the figure two groups of curves are shown, one for D_{in} maximum and the other for D_{in} minimum. Here it must be noted that maximum inner diameter D_{in} is different for all widths. Consider the width, $W = 25$ μm. We can see that the curve with minimum and maximum inner diameters D_{in} meets at $N = 7$. The region enclosed by these two curves cover all possible inductance that can be designed with $W = 25$ μm. It can be seen that the inductance varies from 1 to 10.5 nH and N varies from 1 to 7 with $D_{\text{in}} = 52$ to 375 μm. Similarly, for other widths the region enclosed by the plot with D_{in} maximum and minimum give the possible inductance that can

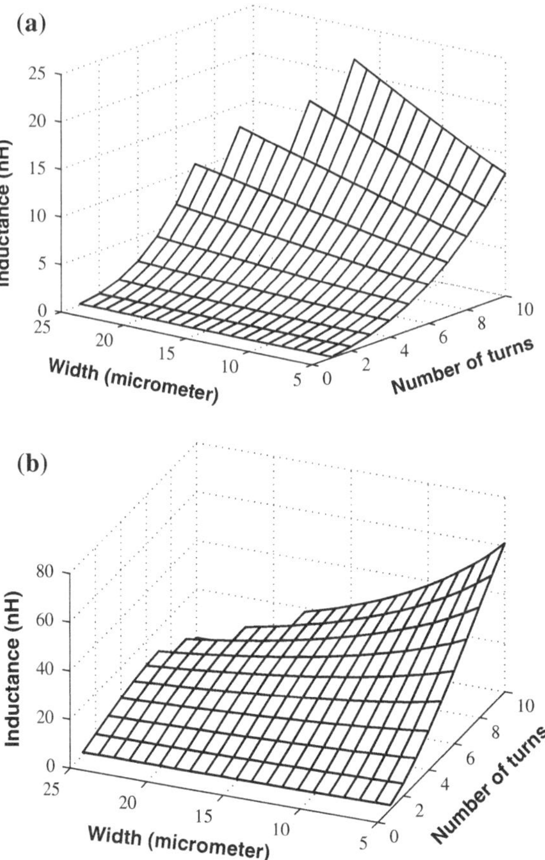

Fig. 2.3 **a** Minimum inductance and, **b** maximum inductance for all combinations of $N = 1$ to 10 and $W = 5$ to 25 μm within the area 400 × 400 μm. Spacing fixed at 2 μm

be designed with each width and the range of turns. Since the graph is shown only for inductance up to 50 nH the intersection point of the plot for $W = 5$ μm is not seen.

A typical problem is to design a fixed inductance. Let us consider that the desired inductance is 10 nH, so we may draw a straight horizontal line of 10 nH. The line cuts the curves of all widths and the corresponding minimum number of turns is 3 and maximum is 9. Moreover, widths $W > 25$ μm will not be able to satisfy the area limit and result in 10 nH inductance. If $W > 25$ μm is to be chosen to realize 10 nH then the area has to be increased. Each point in the graph corresponds to a different inner diameter. Here D_{in} ranges from 52 to 275 μm. Similarly, for L greater than 16 nH, width must be less than 20 μm to satisfy the area limit. In this method of bounding the layout design parameters, the spiral inductor outer diameter is assumed to be given. Since the area is fixed, it can be seen that for some range of inductances, the range of the metal width that can be used becomes limited. So it may be possible that

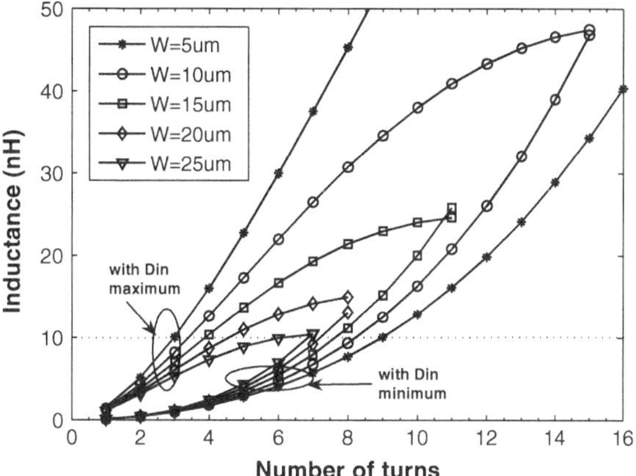

Fig. 2.4 Layout parameter bound curves of possible inductances by varying D_{in} from minimum to maximum for all combinations of number of turns and width that satisfy the area $400 \times 400\,\mu$m. Spacing fixed at $2\,\mu$m

the optimum quality factor obtained using the layout parameter bounding method may be lower as compared to an optimization schedule without any area limitation. However, for a known area, it will always be advantageous to use this method to find the bounds on the width and turns for any inductance and the corresponding inner or outer diameter limits. The feasible region of the optimization is thus identified and the optimum search can be performed within the feasible region only. If we consider the spiral area greater than $400 \times 400\,\mu$m, the size of the envelope will increase as maximum number of turns, N_{max} for each width will increase. Similarly, if the spiral area is less than $400 \times 400\,\mu$m, the envelope size will decrease. Therefore, the bounding curves must be plotted for the maximum inductor area specified. Even for a different area specification, the replotting of the curves would take only a few seconds.

For example, metal width greater that $25\,\mu$m cannot be possibly used to design inductors greater than 10 nH in this area of $400 \times 400\,\mu$m. If width greater than $25\,\mu$m is to be used the area needs to be increased. Because of the fixed area assumption, some possible structures with very large metal widths may not be included. For optimization of inductors at frequencies less than their peak frequency, the quality factor may be lower as compared to an optimization schedule without any area limitation.

The graphical information can be summarized as:

(i) For a specified area the range of inductance values that can be realized by each combination of turn, N and width W is obtained.
(ii) For any desired value of inductance, the bounds on the number of turns, width, and diameter is obtained.

In this way, the bounds on the design parameters are determined and the optimal search can be carried out efficiently. Since the bounding of the design parameters for a large range of inductance values can be done simultaneously, it will shorten the design cycle, especially for applications that require multiple inductors of different values.

2.3 Performance Study of Fixed-Value Inductors Using EM Simulator

The inductance value of a spiral inductor is mainly decided by its geometrical or layout parameters [6]. The performance is determined by its layout parameters and the technological parameters. The first step in the design of spiral inductor involves the finding of the combination of numbers of turns, width, spacing, and inner or outer diameter for a specified inductance value. These layout parameters can be determined from the bounding curve in Sect. 2.2. Depending on the inductor layout and the technology, the associated parasitics due to the ohmic loss in the metal and the losses due to the lossy substrate will vary. To investigate the effects of the parasitics on the performance, a method of moment-based 3D EM simulator is used. For an extensive analysis of the design and performance issues of spiral inductors, the layout parameters are varied for a constant value of inductance [7]. In this way the performance can be compared closely.

2.3.1 Area of the Inductor

Typically, a spiral inductor occupies a huge area on the die. The goal of design has always been to minimize the area since the cost increases proportionately with the area. The area can be reduced by adjusting the inner diameter. But the other goal in the design is also to achieve the desired inductance. Hence, it is important to understand how the area of a desired inductance changes, if we adjust the inner diameter and vary the width and number of turns while the resulting inductance is kept constant. For example, let us consider to design a 10 nH inductor by varying number of turns, width, or spacing and adjust the diameter. The spacing between the metal turns can be kept constant for simplicity. This results in a large number of combinations of the layout parameter and the trend can be clearly depicted if we plot the contour of different areas as the width and number of turns varies as shown in Figs. 2.5 and 2.6. The labels in the contour lines indicate the outer diameter for the corresponding turns and width combination of each 10 nH inductor. The area of the spiral will be $D_{\text{out}} \times D_{\text{out}}$. The outer diameter values gradually decrease along the positive X-axis with the increasing number of turns. Thus if we fix the width and try to reduce the area by decreasing the inner diameter, to achieve the desired inductance the number of turns has to be increased. On the other hand, if we observe the variation of area with width, we can see that the area increases with width. And the area is smallest for the smallest width. If we cross-examine the two figures, we can see

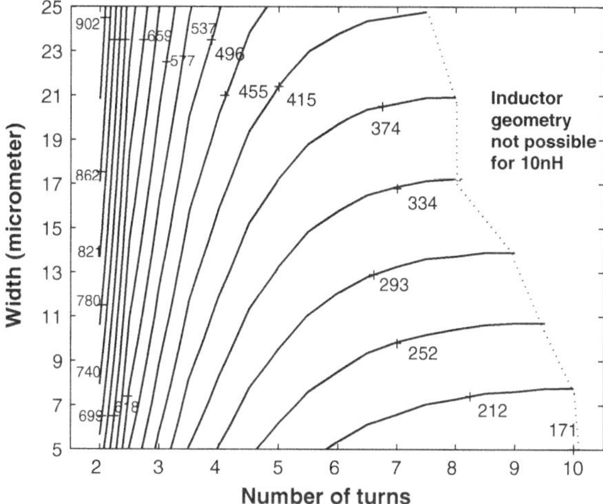

Fig. 2.5 Contour plots of outer diameter (with labels in μm) as a function of width and turns for a 10 nH inductor with spacing 2 μm

that for the same width and turns the inductor with larger spacing occupies a larger area. This is because when the spacing between the tracks is increased, the magnetic coupling decreases. To achieve the same inductance, the inner diameter has to be increased for the same number of turns and width. Thus, the spacing has to be kept as minimum as possible. Therefore, an inductor of a particular inductance with the minimum area can be designed by choosing the smallest width and maximum number of turns combination and keeping the spacing as minimum as possible. However, this smallest inductor is not going to result in the best performance definitely due to the eddy current effect and current crowding in the metal conductor. The performance trend will be discussed in detail in the next subsection. Therefore, a good design would require a careful trading off between the performance and the area occupied by the spiral.

Another method of reducing the area is by stacking the inductors, where two or more individual spiral coils overlap with each other. If the spirals are identical and the mutual coupling factor between the spiral is unity, for an n-layer stack, inductance increases by nearly a factor of n^2 [8]. The same 10 nH inductor design as discussed previously may be done as a two-layer stack of ≈2.5 nH each. In Fig. 2.7, the trend of outer diameter variation is shown for such a design. For the same width and turns combination, the area may be reduced by an average of 53 %. However with stacking, parasitic capacitance increases and hence the self-resonance frequency decreases. The performance of the stack and its planar counterpart is also compared in the next subsection.

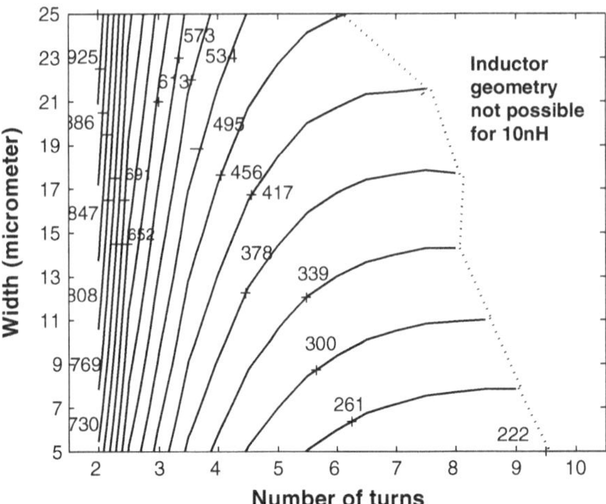

Fig. 2.6 Contour plots of outer diameter (with labels in μm) as a function of width and turns for a 10 nH inductor with spacing 6 μm

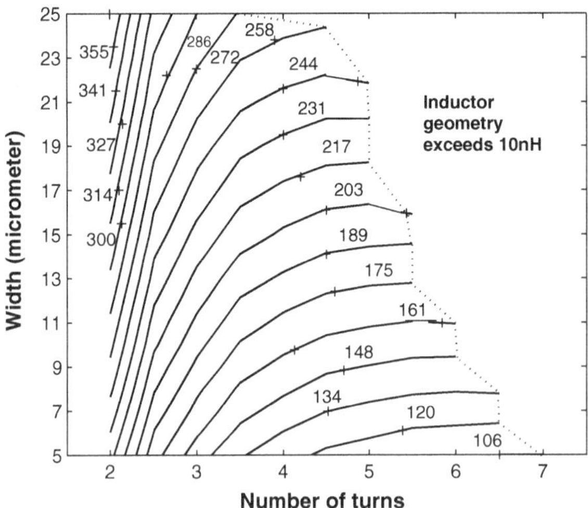

Fig. 2.7 Contour plots of outer diameter (with labels in μm) as a function of width and turns for a 10 nH inductor assumed to be designed by stacking two spirals of 2.5 nH inductance

2.3.2 Quality Factor Variation with the Number of Turns

Quality factor is the most important figure of merit of the inductor. For an inductor, Q is proportional to the energy stored which is equal to the difference between the peak magnetic energy and electric energy. Quality factor of an on-chip spiral inductor increases with frequency and reaches a maximum value after which it decreases due to the ohmic loss in the series resistance and the loss in the substrate. To investigate the effect of varying width, turns, and spacing on quality factor when the inductance is constant, three groups of 10 nH inductors are simulated wherein one of the parameters is varied keeping the other two constant. The layout parameters of these inductors are given in Table 2.1. In Group A the number of turns are varied, in Group B the width is varied, and in Group C the spacing is varied. The structures are simulated using a 3D EM simulator. In simulation, the substrate and the dielectric layers are defined as per the technology parameters of a four-level metal process to reproduce the actual inductor as close as possible. The spiral underpass is in M3. The technological parameters of the design are summarized in Table 2.2. The performance trends are discussed as follows.

The performance of Group A Spiral inductors of 10 nH with turns 3, 4, 5, 6, 7, 8, and 9 were simulated keeping the width and spacing constant at 14 and 2 μm, respectively. Figure 2.8, shows the plot of quality factor for different turns. As expected for a given number of turns, quality factor increases as frequency increases and then decreases due to the parasitics associated. The $Q_{\max}$ depends on the number of turns. As the number of turns increases from 3 to 4, initially $Q_{\max}$ increases but beyond 4, $Q_{\max}$ decreases. The maximum value of $Q_{\max}$ obtained was 6.21 for N=4. The $f_{\max}$ obtained for N=4 also was maximum. As the number of turns increases, keeping the inductance constant, the inner and outer diameters reduce decreasing the area and increasing the total length of the spiral as observed in the previous section. Because of the smaller inner diameter, magnetic fields of the adjacent outer turns will pass through some of the innermost turns, inducing eddy current loops. This will result in nonuniform current in the innermost turns thereby increasing the effective resistance as can be seen from Fig. 2.9, where the real or resistive component of the impedance (Re[Z1]) is plotted for each inductor. As a result, quality factor decreases with increase in the number of turns. In other words, for smaller number of turns since the inner diameter is large, the eddy current effect decreases and the quality factor increases. Thus, spiral inductors designed with larger number of turns to save the area will suffer from low-quality factor. However, further decreasing the number of turns from 4 to 3 by increasing the inner diameter does not improve quality factor but instead increases the area ($D_{\text{out}} \times D_{\text{out}}$) and the total length. The large change in area and length is required as the mutual inductance parameter decreases with less number of positive mutual couplings. With the increase in length, the series resistance of the spiral increases and therefore decreases the quality factor. The area and quality factor can be traded off carefully and the layout parameters of a spiral inductor can be chosen. The f_{res} also increases with the number of turns when the width is fixed since the area is decreased.

2.3.3 Quality Factor Variation with the Metal Width

The width of the metal track is a vital parameter. In Group B, 10 nH spiral inductors of width 6, 8, 10, 12, 14, 16, 18s, and 20 μm were designed while keeping the number

Table 2.1 Layout Parameters

Group name	Number of turns	Width (μm)	Spacing (μm)	D_{in} (μm)	D_{out} (μm)	Total length (μm)	L (nH)	Q_{max}	f_{max} (GHz)	f_{res} (GHz)
A: Different turns	3	14	2	422	514	5600	10.00	6.04	1.06	3.6
	4	14	2	270	394	5296	10.01	6.21	1.06	3.9
	5	14	2	186	342	5264	10.04	6.17	0.94	4
	6	14	2	126	314	5264	10.05	6.04	0.94	4.19
	7	14	2	82	302	5360	10.00	5.84	0.94	4.19
	8	14	2	44	296	5424	10.06	5.67	0.94	4.3
	9	14	2	10	294	5458	10.01	5.58	0.94	4.3
B: Different width	6	6	2	129	221	4192	10.07	5.7	1.71	6.1
	6	8	2	130	246	4502	10.05	5.9	1.38	5.4
	6	10	2	130	270	4788	10.06	6.03	1.22	4.8
	6	12	2	129	293	5050	10.06	6.07	1.06	4.4
	6	14	2	126	314	5264	10.05	6.05	0.94	4.1
	6	16	2	124	336	5502	10.03	5.98	0.87	3.8
	6	18	2	121	357	5716	10.03	5.92	0.82	3.6
	6	20	2	118	378	5930	10.01	5.8	0.75	3.3
	4	14	2	270	394	5296	9.93	6.21	1.06	3.9
C: Different spacing	4	14	6	278	426	5612	10.00	5.8	1.0	3.71
	4	14	10	282	454	5864	10.02	5.5	1.0	3.6
	4	14	14	284	480	6084	10.03	5.3	1.0	3.35

Table 2.2 Technological parameters

Parameter	Values
Substrate resistivity	10 Ω cm
Silicon dielectric constant	11.9
Oxide thickness	4.5 μm
Oxide dielectric constant	4
Conductivity of the metal	5.8×10^5 (Ω cm)$^{-1}$
Metal thickness	1 μm

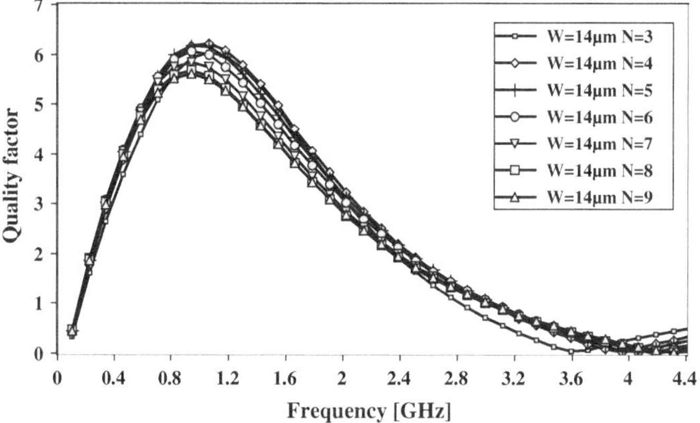

Fig. 2.8 Quality factor for 10 nH inductors designed with number of turns 3, 4, 5, 6, 7, 8, and 9 and the width and spacing fixed at 14 and 2 μm respectively

of turns fixed at 6 and spacing at 2 μm. For a fixed turn, when the same inductance value is realized with larger width, the area increases. In Fig. 2.10, the quality factor for varying width is plotted and it can be seen that as the width increases, the quality factor increases at low frequencies (say at 0.6 GHz). This is because quality factor depends on the series resistance of the metal trace and larger width inductor, which has less resistive loss will have higher quality factor. However, at high frequencies (say 1.8 GHz) quality factor decreases with increase in width. To explain this, the series resistance is plotted against frequency in Fig. 2.11. As frequency increases the resistance increases due to the well-known skin effect and current crowding problem. Skin effects are relatively small below 2 GHz as the metal thickness will be less than the skin depth, nevertheless the current crowding is a strong function of frequency. Current crowding causes an increase in resistance at a much higher rate than the normal linear one especially above a frequency termed as critical frequency [9]. From Fig. 2.11, we can see that for larger width, the current crowding effect begins at lower frequency. Also, with the increase in width, the substrate coupling capacitance increases due to the increase in surface area. Therefore, the quality factor of larger width inductors decays faster and self-resonant frequency also decreases. Both $f_{\max}$ and f_{res} increases for smaller width and depending on the inductor application and the desired operating frequency one can optimize layout parameter appropriately.

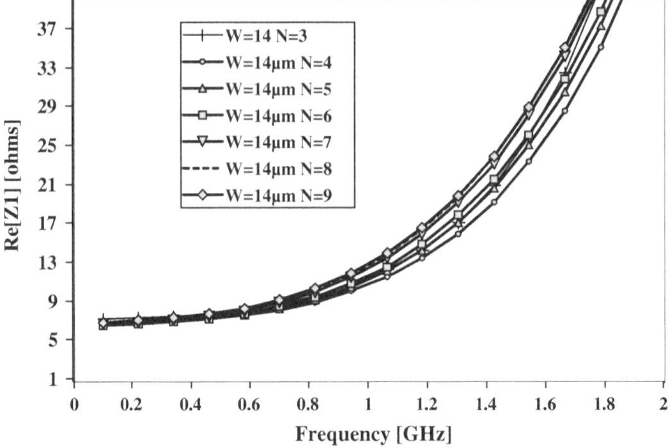

Fig. 2.9 Parasitic series resistance for 10 nH inductors designed with number of turns 3, 4, 5, 6, 7, 8, and 9 and the width and spacing fixed at 14 and 2 μm, respectively

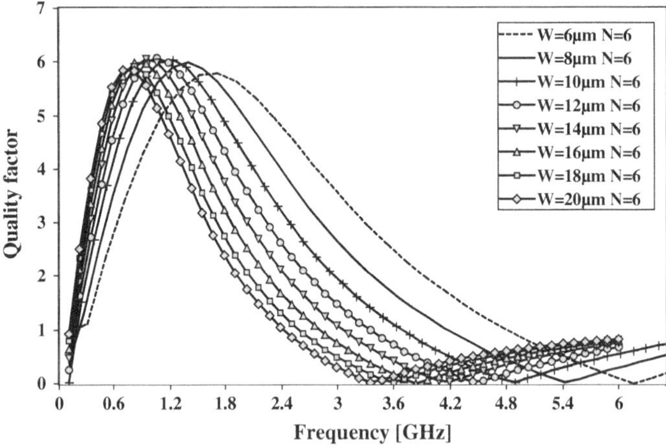

Fig. 2.10 Quality factor for 10 nH inductors designed with different widths of 6, 8, 10, 12, 14, 16, 18, and 20 μm

2.3.4 Quality Factor Variation with the Spacing Between the Metal Tracks

In Group C, the width and the number of turns are kept constant at 14 μm and 4. Four 10 nH inductors of spacing 2, 6, 10, and 14 μm were designed and simulated. In Fig. 2.12, the quality factor for various spacing is compared. When the spacing between the tracks is increased, the magnetic coupling decreases. To achieve the fixed inductance, the inner diameter has to be increased for the same number of

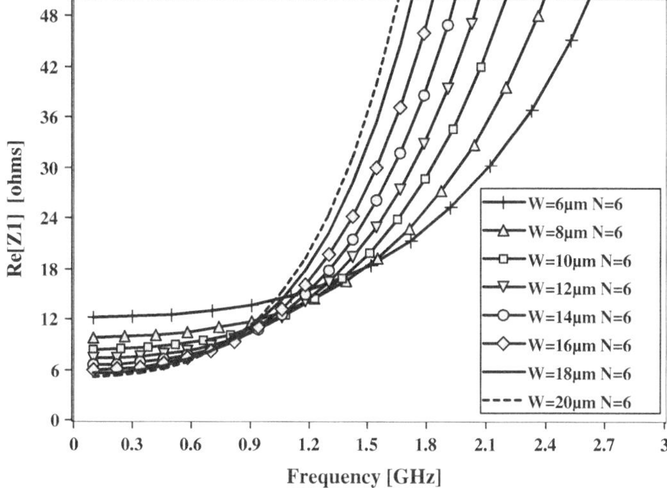

Fig. 2.11 Parasitic series resistance for 10 nH inductors designed with different widths of 6, 8, 10, 12, 14, 16, 18, and 20 µm

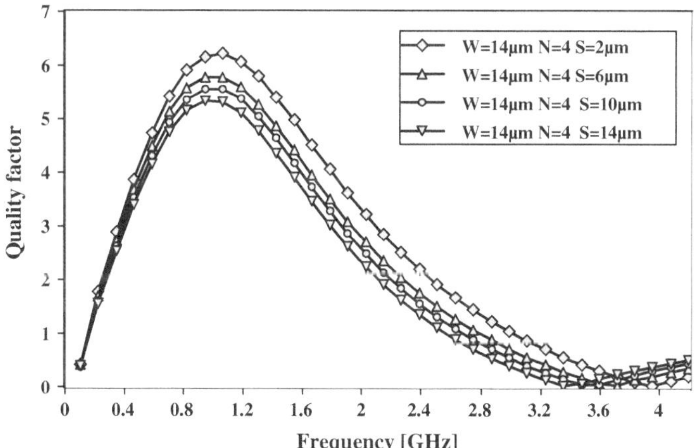

Fig. 2.12 Quality factor for 10 nH inductors designed with different spacing of 2, 6, 10, and 14 µm with number of turns and width fixed at 4 and 14 µm respectively

turns. This increases the total length of the spiral thereby increasing the parasitic series resistance as can be seen from Fig. 2.13, and hence the quality factor is highest for minimum spacing. It can also be seen that the f_{max} and f_{res} also have similar trends as of Q and the best values are obtained for minimum spacing of metal tracks.

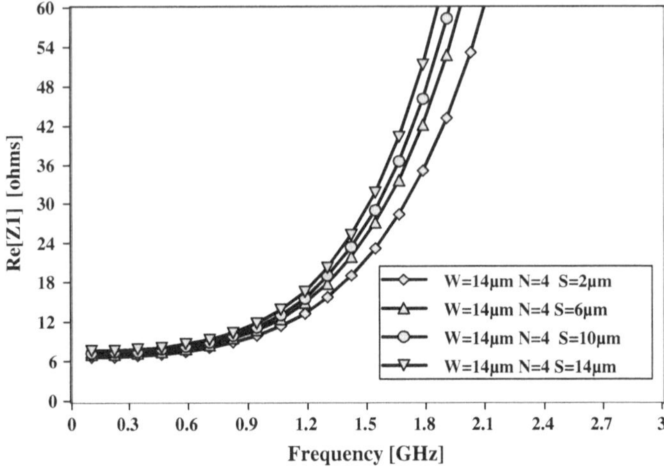

Fig. 2.13 Parasitic series resistance for 10 nH inductors designed with different spacing of 2, 6, 10, and 14 µm

2.4 Efficient Optimization with Bounding of Layout Parameters

To illustrate that the design time and accuracy of a spiral inductor optimization schedule is improved using the bounding curves an enumeration type optimization algorithm similar to [10] is implemented and the lower and upper bounds on the constraints of the design parameters are given according to the bounding curves. Since only the possible combinations of width and number of turns that will result in the desired value of inductance is given, the step to check whether design exists is not required as in [10]. The steps of the optimization algorithm are summarized below:

(i) Input the design specifications, such as the desired inductance value, technology parameters, and specified operating frequency.
(ii) For $L = L_{desired}$ refer the layout parameter bounds diagram and read the range of the number of turns and width that will result in the exact value of desired inductance. Assign $N = N_{min}$ to N_{max} and $W = W_{min}$ to W_{max}.
(iii) For each N and W combination adjust the inner diameter, D_{in} is so that $L = L_{desired}$ and calculate the total length of the spiral.
(iv) Compute the quality factor for each combination of turns and width at the desired operating frequency using the lumped element model [11] and store it.
(v) The maximum quality factor Q_{max} is the optimum solution and its corresponding layout parameters are the optimum layout parameters.
(vi) Verify the design using a 3D electromagnetic simulator.

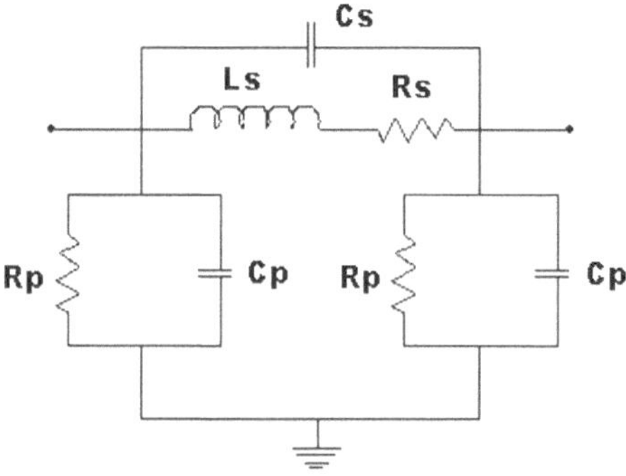

Fig. 2.14 Simplified lumped element model of on-chip spiral inductor on silicon

2.4.1 Lumped Element Model of a Planar Spiral Inductor

The optimization is based on the well-accepted accurate physical model [11] shown in Fig. 2.14. The lumped element model consists of the inductor and its associated parasitics. The model gives the equivalent circuit representation of the inductor which can be used to characterize the electrical behavior of the component. The realization of an ideal lumped element is impossible and therefore a model should account for the frequency-dependent characteristics resulting from the fringing field, proximity effects, substrate material, conductor thickness, etc. The model may be valid upto the self-resonance frequency of the inductor. The inductance and resistance of the spiral and underpass are represented by the series inductance, L_s and series resistance R_s respectively. The capacitive coupling due to the crosstalk between the adjacent turns and the overlap between the spiral and the underpass is modeled by C_s. C_p and R_p represents the overall parasitic effect of oxide and Si substrate. L_s is calculated similarly as discussed before. The parameters R_s, C_s, C_p and R_p are calculated as

$$R_s = \frac{\rho\, l}{W\, \delta (1 - e^{t/\delta})} \tag{2.9}$$

$$C_s = n\, W^2\, \frac{\varepsilon_{ox}}{t'_{ox}} \tag{2.10}$$

$$C_p = C_{ox}\, \frac{1 + \omega^2\, (C_{ox} + C_{Si})\, C_{Si}\, R_{Si}^2}{1 + \omega^2\, (C_{ox} + C_{Si})^2\, R_{Si}^2} \tag{2.11}$$

$$R_p = \frac{1}{\omega^2\, C_{ox}^2\, R_{Si}} + \frac{R_{Si}\, (C_{ox} + C_{Si})^2}{C_{ox}^2} \tag{2.12}$$

where ρ is the resistivity, l is the total spiral length, W is the width, and δ is the skin depth, t is the physical thickness of the metal, n is the number of overlaps, t'_{ox} is the oxide thickness between the spiral and the underpass. C_{Si} and R_{Si} are the capacitance and resistance of the silicon substrate, and C_{ox} is the oxide capacitance between the spiral and the silicon substrate calculated as

$$C_{ox} = \frac{1}{2} l\ W\ \frac{\varepsilon_{ox}}{t_{ox}} \tag{2.13}$$

$$R_{Si} = \frac{2}{l\ W\ G_{\text{sub}}} \tag{2.14}$$

$$C_{Si} = \frac{1}{2} l\ W\ C_{\text{sub}} \tag{2.15}$$

where G_{sub} and C_{sub} are the conductance and capacitance per unit area of the silicon substrate and t_{ox} is thickness of the oxide layer separating the spiral and the substrate.

2.4.2 Calculation of Figure of Merits

Quality factor is proportional to the magnetic energy stored which is equal to the difference between the peak magnetic energy and electric energy.

$$Q = 2\pi\ \frac{\text{peak magnetic energy} - \text{peak electric energy}}{\text{energy loss in one cycle}} \tag{2.16}$$

Based on this definition Q is calculated as [12]

$$Q = \frac{\omega L_s}{R_s} \frac{R_p}{R_p + \left[(\frac{\omega L_s}{R_s})^2 + 1 \right] R_s} \times \left[1 - \frac{R_s^2\ (C_s + C_p)}{L_s} - \omega^2 L_s (C_s + C_p) \right] \tag{2.17}$$

where L_s, R_s, C_s, R_p, and C_p are model parameters defined above. The self-resonant frequency, f_{res} of an inductor is the frequency where the inductive reactance and the parasitic capacitive reactance become equal and opposite in sign. This is determined by the frequency point where Q becomes zero. The optimum frequency, $f_{\max}$ is the frequency at which Q is maximum.

2.4.3 Performance Evaluation with an Optimization Example

In this section we demonstrate the optimization methodology by taking up a problem to optimize the design of 6 nH inductor at 2 GHz. The design constraints are given

Table 2.3 Optimization constraints

Parameter	Values
Desired inductance	6 nH
Operating frequency	2 GHz
Outer diameter	≤ 400 μm

in Table 2.3. The optimization is performed with the same technology parameters in Table 2.2. A tolerance of 2 % is allowed on the inductance value. For an inductance of 6 nH, from the bound curves in Fig. 2.4 (Sect. 2), we determine the upper and lower bounds on the number of turns and width. The number of turns can vary from 3 to 7 for W varying from 5 to 25 μm. The quality factors at 2 GHz as a function of varying width and turns are plotted in Fig. 2.15 and the corresponding outer diameter is shown in Fig. 2.16. The highest value of Q is 7.13 for $W = 12$ μm and $N = 4.5$ and is marked by a circle. The inner diameter (D_{in}) is 133 μm and outer diameter (D_{out}) is 255 μm. We verified the predicted inductance and the quality factor using a 3D electromagnetic simulator [13]. The frequency dependence of inductance and quality factor for this optimum design are plotted in Figs. 2.17 and 2.18 respectively. The inductance calculated using Greenhouse method [2] is 5.92 nH, however, at 2 GHz from Fig. 2.17 the effective inductance is 6.34 nH. There is an error of 6.62 % only. In general, inductors are used only in their inductive region, i.e, the useful band of operation of an integrated inductor [6] where the inductance value remains relatively constant.

Spiral inductors consume a lot of die area in RF circuitry as compared to the area required by active devices. To minimize the cost, the performance can be carefully traded off with the area ($D_{out} \times D_{out}$). These tradeoffs can also be explored from Figs. 2.15 and 2.16. Inductors with larger number of turns have smaller area but quality factor is lower because of smaller inner diameter. The magnetic fields of the

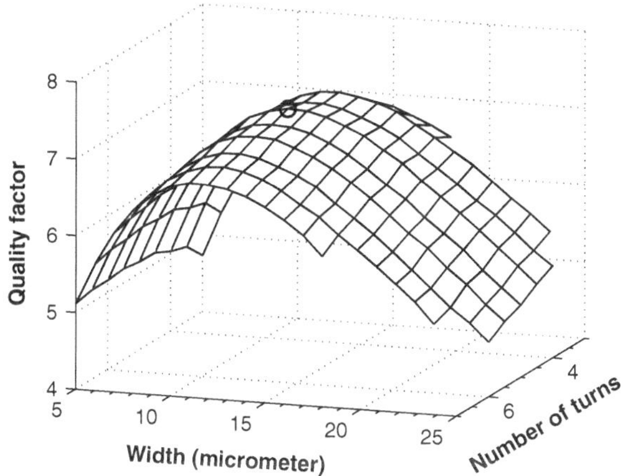

Fig. 2.15 Quality factors for 6 nH inductors as a function of width and number of turns

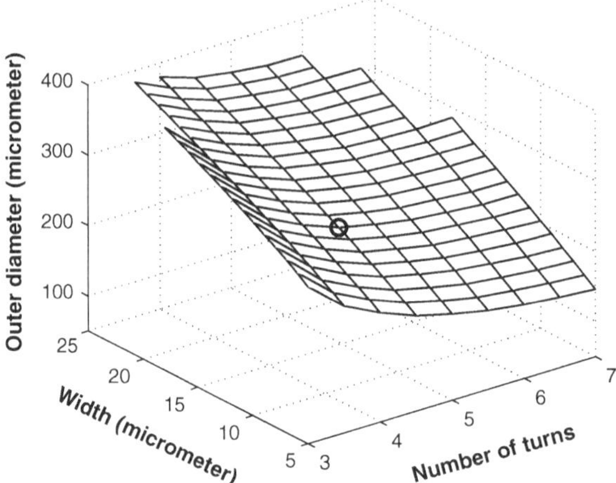

Fig. 2.16 Outer diameters for 6 nH inductors as a function of width and number of turns

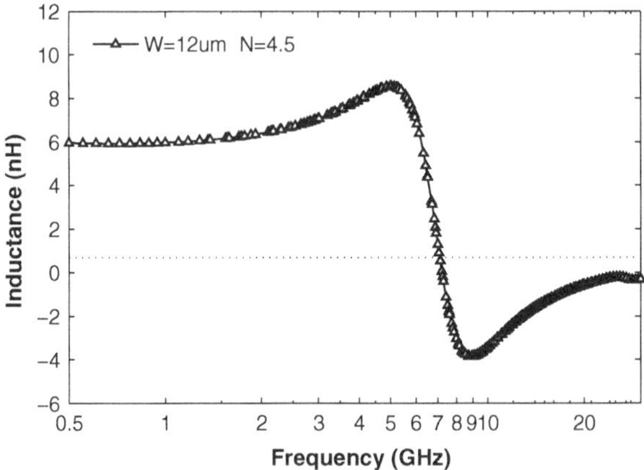

Fig. 2.17 Inductance of the optimum design of 6 nH inductor as a function of frequency

adjacent outer turns will pass through some of the innermost turns, inducing eddy current loops which result in non-uniform current in the innermost turns thereby increasing the effective resistance and hence lowering the quality factor. This eddy current effect can be minimized by increasing the inner diameter and realized with the same inductance of 2–3 turns, but the area will also increase. However, it does not improve quality factor due to the increase in the series resistance as the total length increases with increase in area. For a fixed turn the spiral area also increases with the increase in width. Spiral structure may be selected considering both the quality factor and area. For a 5 % reduction in the quality factor, area can be saved by 39 % as compared to the optimum structure with the combination $W = 9\,\mu m$, $N = 6$ and $D_{out} = 199\,\mu m$ that results $Q = 6.7$. Similarly for a 10 % reduction in the quality

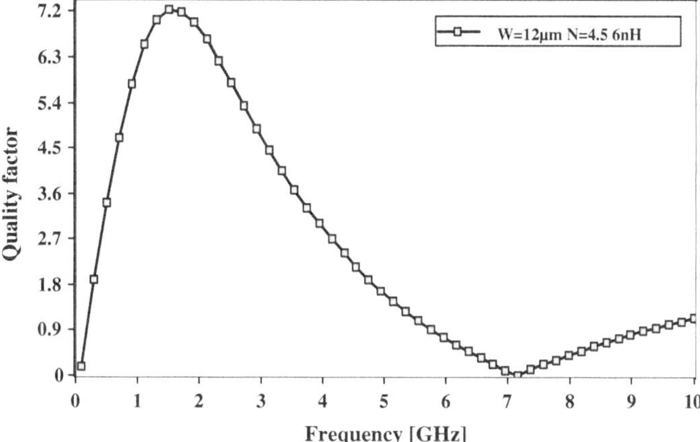

Fig. 2.18 Quality factor of the optimum design of 6 nH inductor as a function of frequency

factor, area can be saved by 49 % as compared to the optimum structure with the combination $W = 8\,\mu m$, $N = 7$ and $D_{out} = 181\,\mu m$ which results $Q = 6.4$.

In the literature, spiral inductor optimization techniques are presented for different process parameters at different operating frequencies. Hence, it would be difficult to compare the results closely. For a fair comparison we have repeated the proposed optimization method using the process parameters employed in [14–16]. The spacing was fixed at 2 μm since the turn-to-turn spacing in the published results was 2 μm. The comparison of the proposed method with other optimization techniques [14–16] for inductance values close to 6 nH is given in Table 2.4. Enumeration method always results in a global optimum solution as compared to numerical algorithms that may sometimes lead to non-convergence and local optimum solutions.

We have seen that the bounding of the layout parameters was performed based on the well-accepted inductance calculation algorithm developed by Greenhouse and in the optimization algorithm presented, a lumped element model of the spiral inductor was used in which the inductance was also calculated using the same formula. Optimization algorithms published in the literature are also based on this model. Since the scalable inductor model has shown good agreement with measured and published data the bounding of the layout parameter algorithm also results in the range of the design parameters that results in the desired inductance values. In Table 2.5, the verification results of the bounding method with some of the structures published are presented. The layout parameters and the inductance values match with the bounding curve of Fig. 2.4. Since the parameter bounds determined are a wide range that will result in the desired inductance values, the bounding will have negligible error.

Table 2.4 Performance comparison of optimization techniques

Methods	L (nH)	Process parameters					Optimized layout			Run
		t_{ox} (μm)	t_{metal} (μm)	S (μm)	freq (GHz)	Q_{max}	W (μm)	N	D_{out} (μm)	time (s)
Hershenson [14]	6	5.2	0.9 $\sigma_M = 3 \times 10^5$	1.9	2.5	4	7.8	4.75	206.3	≤ 1
	6	5.2	0.9 $\sigma_M = 3 \times 10^5$ with PGS	1.9	1.5	4.2	13	3.75	292.2	
Zhan [15]	5.7	5	1 $R_{sheet} = 20$	2	2	10.78	17.5	3	400	41
Nieuwoudt [16]	6	6	0.5 $\sigma_M = 5.8 \times 10^5$, $\sigma_{sub} = 7000$	0.5	0.9	3.55	27.26	4.45	367.5	$\cong 14$
Proposed	6	5.2	0.9 $\sigma_M = 3 \times 10^5$, $\sigma_{sub} = 10$	1.9	2.5	5.21	13	3.5	308	0.219
	6	6	0.5 $\sigma_M = 5.8 \times 10^5$, $\sigma_{sub} = 7000$	0.5	0.9	4.60	30	3.5	432	
	6	4.5	1 $\sigma_M = 5.8 \times 10^5$, $\sigma_{sub} = 10$	2	2	7.13	12	4.5	255	

σ_M: conductivity of the metal $(\Omega\ \text{cm})^{-1}$; PGS: protective ground sheild
σ_{sub}: conductivity of the substrate $(\Omega\ \text{m})^{-1}$; R_{sheet}: sheet resistance of the metal (m Ω/□)

Table 2.5 Verification of layout parameters

Ref.	Number of turns	Width (μm)	Spacing (μm)	D_{out} (μm)	Measured L (nH)
[17]	3.75	13	1.9	292	6
[17]	5.75	10	1.9	339	16.2
[18]	9.25	5.2	2	145	6
[19]	4	18	2	346	5.9
[19]	5	18	2	346	7.5

2.4.4 Computational Speed

The optimization of 6 nH inductor discussed before is completed in 0.219 s of CPU time using the simple and accurate expression [20] for inductance calculation and 16.81 s of CPU time using Greenhouse method [2]. In enumeration method the time required for the optimization or the number of function evaluations will depend on the discretization of the design space (N, W and D_{out}). In our design example, we have chosen $11 \times 21 \times 98$ grid. The W and D_{out} was incremented by 1, 4 μm, respectively, and N was incremented by half-turns each time. These step sizes were chosen with assurance that optimum design was not missed out. The optimization method requires a total function evaluation of 5,393. Since the quality factor was calculated only at the combinations which result in $L = 6$ nH the quality factor function evaluation is only 175. But an enumeration method without layout parameter bounding and with the same design constraints would require a total function evaluation of 37,044. This will again increase for arbitrarily decided constraints and may require as large as a million function evaluations [16]. Therefore with layout parameter bounding, a large proportion of the sample points that are redundant for a desired inductance value is pruned off and the number of function evaluations is reduced significantly.

In Table 2.4, a comparison of the computation time is also included. Geometric programming (GP) takes the minimum time of less than a second, among the numerical methods. The computation time of our proposed method is comparable to GP but less than other methods. Also, the global optimization of inductance range 1–20 nH as discussed before, is completed in 6.40 s of CPU time. To compare the computation time with GP more closely, we have also implemented geometric programming algorithm [14]. For the same inductance range, with the same technological parameters, geometric programming performs the optimization in 7.36 s of CPU time. In Fig. 2.19, the global optimal tradeoff comparison at 2.4 GHz is shown. Moreover if a field solver, which requires an average simulation time of 5 min per frequency point, is used to get the same result it may take several days. Therefore, with layout parameter bounding, the computation time of an enumeration method is even less than or comparable to other numerical algorithms of [14–16].

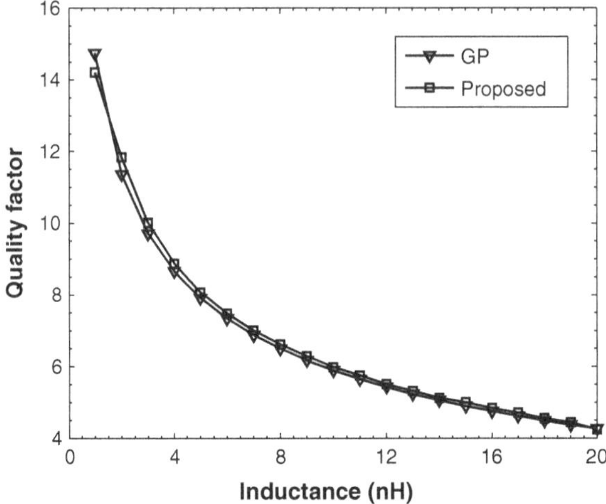

Fig. 2.19 Comparison of global optimal tradeoff curves for inductance range 1–20 nH at 2.4 GHz (GP-geometric programming)

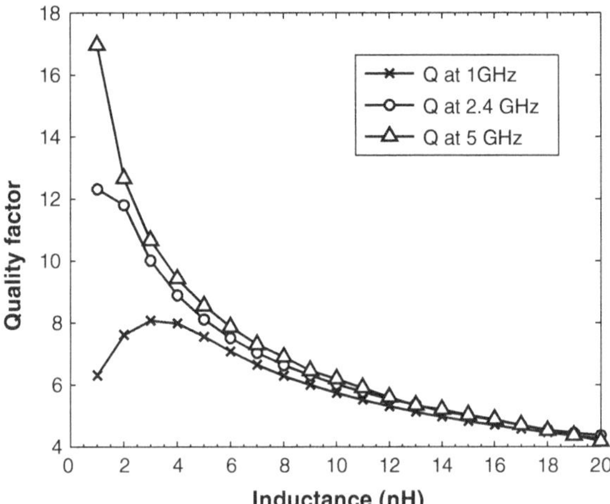

Fig. 2.20 Global optimal quality factor and inductance trade off curves at 1, 2.4 and 5 GHz

2.4.5 Global Optimal Quality Factor Tradeoff Curve

In the previous section we discussed the optimization method considering the optimization of 6 nH at 2 GHz as an example. We repeated the optimization to generate the global optimal tradeoff curves for inductance range 1–20 nH at 1 Hz, 2.4, and 5 GHz as shown in Fig. 2.20. The major inference is that the quality factor decreases with the increase in inductance. It may appear that in the quality factor plot for 1 GHz, Q does not increase with inductance for an inductor value less than 4 nH. This is because of the specific design constraint on layout parameters. It may be

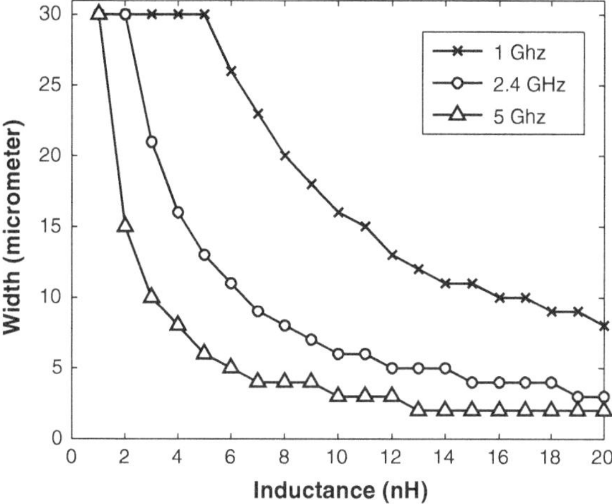

Fig. 2.21 Optimum width versus inductance at 1, 2.4 and 5 GHz

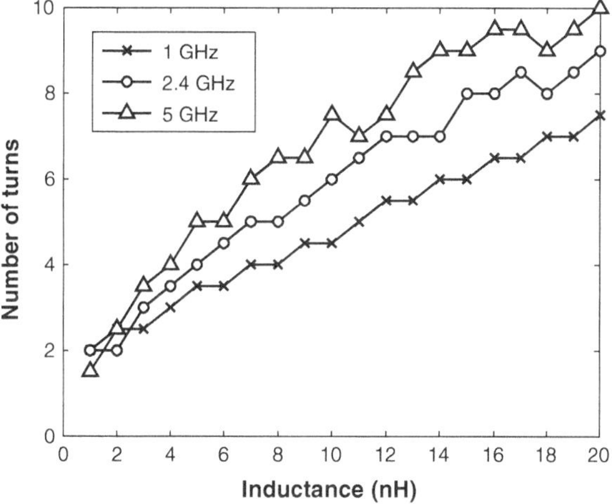

Fig. 2.22 Optimum number of turns versus inductance at 1, 2.4 and 5 GHz

noted that at 1 GHz it is still possible to get higher Q for inductances less than 4 nH if one increases the limits of the design constraints of the optimization. Ideally, the quality factor will increase with the inductance if there is no limitation on the design constraint. The trend of variation of the corresponding optimum width, number of turns, and outer diameter are plotted in Figs. 2.21, 2.22 and 2.23 respectively. We can see that optimum width decreases with inductance while the number of turns increases in all the three cases. Also, for any inductance, as the frequency increases the optimum width decreases and the number of turns increases. These curves gives a good overview of inductance values and their quality factor at different frequencies for on-chip inductors and one can quickly estimate what values are appropriate for the desired application.

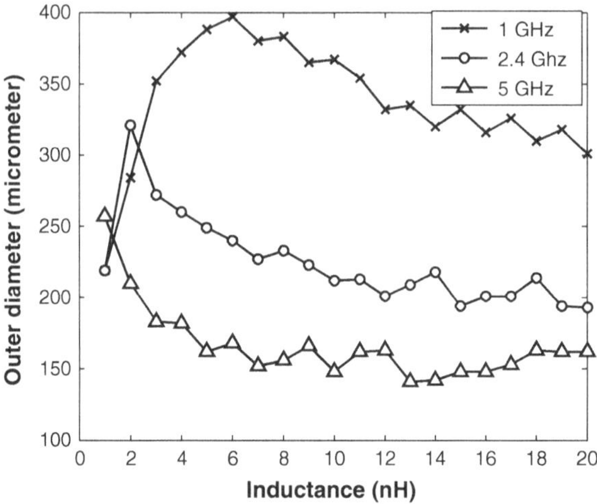

Fig. 2.23 Optimum outer diameter versus inductance at 1, 2.4 and 5 GHz

2.4.6 Peak Quality Factor Variation with Inductance

For any inductance, the peak quality factor increases with the increase in frequency as we vary the layout parameters until it reaches its maximum peak quality factor, Q_{max} and the corresponding frequency is referred as f_{max}. Beyond this frequency where the highest value of quality factor is obtained, the peak quality factor will begin to decrease as we change the layout parameters. There also exists a tradeoff of the maximum quality factor, Q_{max} and the frequency at which it occurs, f_{max}. Hence we have optimized the quality factor without the frequency constraint to find Q_{max} and f_{max} for inductance range 1–20 nH. The result is shown in Fig. 2.24. It gives the Q_{max} and f_{max} pair for each inductance value. This indicates that under the specified design constraint, for example, an inductance of 5 nH can be designed with the maximum quality factor of 8.5 at a frequency of 6.5 GHz. The optimum combination of N, W and D_{out} is given in Table 2.6. The results presented in Table 2.6 being computed using a lumped element model, upon verification with a 3D electromagnetic simulator may result in slight error similar to that mentioned before for our design example of 6nH at 2 GHz. However, the error is within the manufacturing tolerance of spiral inductor realization [16].

2.5 Optimization Using EM Simulator

The analysis of the performance trend of a fixed inductance value with the layout parameters in the previous section has established the basic insights necessary for an inductor optimization. With this knowledge the optimized layout can be identified

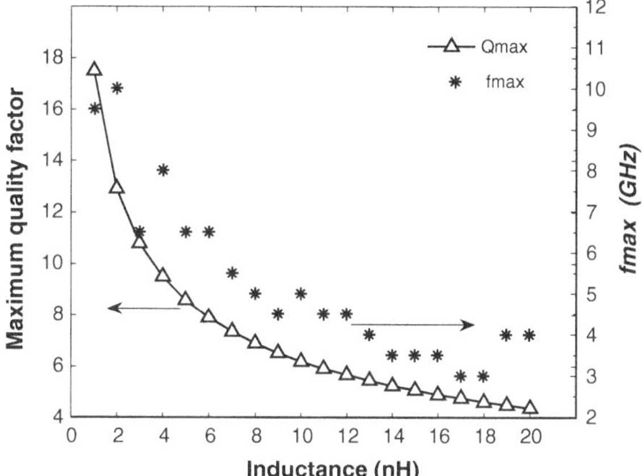

Fig. 2.24 Maximum quality factor and their corresponding peak frequency for inductance range 1–20 nH

by simulating only few structures using an EM simulator to get the accurate design and save design time. We propose here an algorithm which consists of the following minimum steps required to design and optimize a spiral inductor by simulating few inductor structures using a 3D EM simulator for a given technology.

(i) Generate all the possible combinations of the layout parameters for the desired inductance value by varying the width and turns and adjusting the inner diameter. Spacing must be kept at a minimum as large spacing increases the area and degrades the performance.
(ii) For the desired operating frequency, identify the optimum width for the desired inductance value to get the highest quality factor by simulating a few structures of the same number of turns but different widths as demonstrated by results in Fig. 2.10 in the previous section. The $Q_{\max}$ and $f_{\max}$ strongly depends on the width of the spiral. At the desired operating frequency, the width of the structure with the highest quality factor is the optimum width. If there is more than one optimum width, choose the larger one for smaller series resistance or choose the smaller one for higher self-resonating frequency.
(iii) Simulate a group of inductors of optimum width but different turns as demonstrated by results in Fig. 2.8 in previous section. The structure with the highest quality factor is the optimized inductor. An examination of the tradeoff between the quality factor and area can also be carried out. Varying the number of turns, $f_{\max}$ remains constant with a slight variation in $Q_{\max}$ and f_{res}. Since inductors occupy considerable area on a chip, the area minimization is important.

For example, we consider that an inductor of 10 nH is to be optimized for a desired operating frequency of 1 GHz. We generate all the possible layout combinations for width varying from 5–20 μm with a step of 1 μm and number of turns varying from 2–10 with a step of 0.5 turn. The step size is chosen with certainty that the optimum

Table 2.6 Maximum quality factor and the corresponding optimum layout parameters

Inductance (nH)	Width (μm)	Turns	D_{out} (μm)	Q_{max}	f_{max} (GHz)
1	17	2	169	17.4	9.5
2	8	3	147	12.9	10.0
3	8	3.5	168	10.7	6.5
4	5	4.5	143	9.4	8.0
5	5	5	152	8.5	6.5
6	4	5.5	147	7.8	6.5
7	4	6	152	7.3	5.5
8	4	6.5	156	6.8	5.0
9	4	6.5	166	6.5	4.5
10	3	7.5	148	6.1	5.0
11	3	7	162	5.9	4.5
12	3	7.5	163	5.6	4.5
13	3	8	164	5.4	4.0
14	3	8.5	165	5.2	3.5
15	3	8.5	171	5.0	3.5
16	3	8.5	177	4.9	3.5
17	3	9	177	4.7	3.0
18	3	8.5	189	4.6	3.0
19	2	9.5	162	4.5	4.0
20	2	10	162	4.3	4.0

design is not missed out. Spacing is kept constant at 2 μm. We get a total of 227 structures with different combinations. The relevant combinations are reduced to 15 numbers based on the algorithm proposed above. When they are simulated, one gets an optimum width of 12 or 14 μm. We choose here 14 μm as a larger width will have smaller series resistance. We get the optimum 10 nH inductor with quality factor of 5.9 at 1 GHz for the same technology in Table 2.2. The number of turns is 4 and the area is 394 × 394 μm^2. The performance may be traded off with the area. With a 5 % decrease in the quality factor we may choose an inductor with 7 turns as the optimum one, saving an area by 41.2 %. Similarly, we have optimized 1 and 6 nH inductors at 5 and 2 GHz respectively. For 1 nH, we obtained an optimum width of 16 μm at 5 GHz and optimum turn of 2. The highest quality factor obtained is 12 with an area of 182 × 182 μm^2. For 6 nH, the optimum width is 10 μm at 2 GHz. The highest quality factor obtained is 6.74 for 5 turns and area of 226 × 226 μm^2. The optimum layout was obtained by simulating only 12 and 15 structures of a total of 117 and 200 possible structures for 1 and 6 nH respectively.

In the literature, spiral inductor optimizations are reported with different technology and a fair comparison would be difficult. The advantage of the proposed methodology becomes very clear when the design is compared with lumped element-based method. We compare this result with the previous optimization based on lumped element model in the same technology. The comparison is given in Table 2.7. As expected, the lumped element model overestimates the performance; however, the

Table 2.7 Comparison of optimized inductors

L (nH)	Freq (GHz)	Lumped element model-based simulation				Electromagnetic simulation			
		Q	W (μm)	N	D_{out} (μm)	Q	W (μm)	N	D_{out} (μm)
1	5	15.18	20	1.4	234	12	16	2	182
6	2	7.94	12	4	279	6.7	10	5	226
10	1	5.45	16	5	362	5.9	14	4	394

most striking difference is that both resulted in different optimized geometries. For example, in the case of 1 nH the lumped element simulation gives an optimum width of 20 μm as compared to 16 μm by EM simulation. The lumped element method does not estimate the influence of substrate parasitics and frequency dependence accurately. Moreover, an optimization based on it always needs to be verified. The EM simulation results are more accurate as compared to lumped element-based methods. Therefore, for a given technology, the optimum combination of the number of turns, width, and inner diameter that results in the highest quality factor at the desired operating frequency can be confidently determined for any value of inductance using a 3D simulator by the proposed method. It is to be noted that this optimization method is computationally not very expensive since less than 10 % of the possible structures needs to be simulated to finalize the optimum one as illustrated. From these illustrations, the advantage of the performance trend study keeping the inductance value fixed is evident. This method ultimately results in the optimized design with simulation of less than 10 % of the possible combinations.

2.6 Summary

We have developed an efficient method for bounding the layout design parameters of on-chip spiral inductors. The bounding algorithm results in several curves for various widths as a function of the number of turns and the selection of upper and lower bounds of optimization variables was done graphically. With bounding curves the feasible region of optimization of a large inductance range that satisfies the same area specification was identified. We have also demonstrated the importance and advantages of studying the performance tradeoffs of the spiral inductor, keeping the inductance constant. The metal track width must be optimized for the desired operating frequency since f_{max} is a strong function of width. With the number of turns, f_{max} remains almost constant and quality factor changes slightly, therefore the number of turns must be selected to optimize the area. Based on these insights obtained from the performance trends of a fixed value inductor design, the optimized layout was quickly determined using only few EM simulations of inductor structures as illustrated by the optimization of 1, 6, and 10 nH at 5, 2 and 1 GHz respectively.

An enumeration optimization algorithm was implemented based on layout parameter bounding. The number of function evaluations was significantly reduced and optimization took less than 1 s of CPU time. The results of optimization were also verified using a 3D EM simulator. Since the feasible region for any desired inductance value is determined apriori the optimization results in global solution and the method is very fast. Since bounding curves can be tailored to include all the desired range of inductance the method is more advantageous when multiple inductors of different values are to be optimized. Several important fundamental tradeoffs of the design such as quality factor and area, quality factor and inductance, quality factor and operating frequency, maximum quality factor and the peak frequency, etc., for inductance values ranging from 1 to 20 nH were explored in a few seconds. With layout parameter bounding, enumeration method is proved to be as fast as other numerical algorithms. Enumeration method always results in a global optimum solution. Hence with layout parameter bounding, optimum spiral inductors can be synthesized and analyzed in an easy and simple manner in a few seconds.

References

1. Haobijam, G., Paily, R.: Efficient optimization of integrated spiral inductor with bounding of layout design parameters. Analog Integr. Circ. Sig. Process. **51**(3), 131–140 (June 2007)
2. Greenhouse, H.M.: Design of planar rectangular microelectronic inductors. IEEE Trans. Parts Hybrids Packag **10**(2), 101–109 (1974)
3. Grover, F.W.: Inductance calculations. Van Nostrand, Princeton, New Jersey (1946), (reprinted by Dover Publications 1962)
4. Long, J.R., Copeland, M.A.: The modeling, characterization, and design of monolithic inductors for silicon RF IC's. IEEE J. Solid State Circ. **32**(3), 357–369 (1997)
5. Sia, C.B., Hong, B.H., Chan, K.W., Yeo, K.S., Ma, J.G., Do, M.A.: Physical layout design optimization of integrated spiral inductors for silicon-based RFIC. IEEE Trans. Electr. Devices **52**(12), 2559–2567 (2005)
6. Koutsoyannopoulos, Y.K., Papananos, Y.: Systematic analysis and modeling of integrated inductors and transformers in RFIC design. IEEE Trans. Circ. Syst. II **47**(8), 699–713 (2000)
7. Haobijam, G., Paily, R.: Performance study of fixed value inductors and their optimization using electromagnetic simulator. Microwave Opt. Technol. Lett. **50**(5), 1205–1210 (2008)
8. Zolfaghari, A., Chan, A., Razavi, B.: Stacked inductors and transformers in CMOS technology. IEEE J. Solid State Circ. **36**(4), 620–628 (2001)
9. Kuhn, W.B., Ibrahim, N.: Analysis of current crowding effects in multiturn spiral inductors. IEEE Trans. Microw. Theory Tech. **49**(1), 31–38 (2001)
10. Post, J.E.: Optimizing the design of spiral inductors on silicon. EEE Trans. Circ. Syst. II: Analog Digit. Sig. Process. **47**(1), 15–17 (2000)
11. Yue, C.P., Wong, S.S.: Physical modeling of spiral inductors on silicon. IEEE Trans. Electr. Devices **47**(3), 560–568 (2000)
12. Yue, C.P., Wong, S.S.: On-chip spiral inductors with patterned ground shields for Si-based RF IC's. IEEE J. Solid State Circ. **33**(5), 743–752 (1998)
13. Farina, M., Rozzi, T.: A 3-D integral equation-based approach to the analysis of real-life MMICs-application to microelectromechanical systems. IEEE Trans. Microw. Theory Tech. **49**(12), 2235–2240 (2001)
14. Hershenson, M.M., Mohan, S.S., Boyd, S.P., Lee, T.H.: Optimization of inductor circuits via geometric programming. In: Proceedings of the Design Automation Conference, pp. 994–998 (1999)

15. Zhan, Y., Sapatnekar, S.S.: Optimization of integrated spiral inductors using sequential quadratic programming. In: Proceedings of the IEEE Design, Automation and Test in Europe Conference and Exhibition, vol. 1, pp. 622–627 (2004)
16. Nieuwoudt, A., Massoud, Y.: Variability-aware multilevel integrated spiral inductor synthesis. IEEE Trans. Comput. Aided Design Integr. Circ. Syst. **25**(12), 2613–2625 (2006)
17. Pettenpaul, E., Kapusta, H., Weisberger, A., Mampe, H., Luginsland, J., Wolff, I.: Cad models of lumped elements on gaas up to 18 ghz. IEEE Trans. Microwave Theory Tech. **36**, 294–304 (1988)
18. Hershenson, M., Lee, T.H., Mohan, S.S., Yue, C.P., Wong, S.S.: Modeling and characterization of on-chip transformers. In Proceedings of the IEEE International Electron Devices Meeting, pp. 531–534 (1998)
19. Normak, U.: Integrated transformer. KTH Kista, Kista (1998)
20. Mohan, S., Hershenson, M., Boyd, S., Lee, T.: Simple accurate expressions for planar spiral inductances. IEEE J. Solid State Circ. **34**(10), 1419–1424 (1999)

Chapter 3
Multilayer Pyramidal Symmetric Inductor

3.1 Introduction

In most of the integrated circuits like amplifiers, mixers, oscillators, etc., the differential topology is preferred because of its less sensitivity to noise and interference. There are mainly two categories of differential inductor design found in the literature. The first one is a pair of asymmetric planar inductors connected together in series [1] as shown in Fig. 1.9 to make it symmetric (differential). Since the currents always flow in opposite direction in these two inductors, there must be enough spacing between them to minimize electromagnetic coupling. As a result, the overall area occupied is very large. The second one is the planar symmetric inductor of [2] as shown in Fig. 1.10 which is realized by joining coupled microstrip from one side of an axis of symmetry to the other using a number of cross-over and cross-under connections. An intermediate metal layer is dedicated for the underpass of the cross-coupled connections. The center tapped idea was proposed in [3] for balanced circuits and this type of winding of the metal trace was first applied to monolithic transformers [4]. The symmetrical inductor under differential excitation results in a higher quality factor and self resonance frequency. It also occupies less area than its equivalent pair of asymmetrical inductors. Since these structures are planar, the area is still large. Minimization of inductor area is equally important as enhancing the performance to reduce the production cost. A multilevel symmetric inductor can be realized by stacking two differential inductor of [2] as shown in Fig. 3.1. The structure is a natural extension of the planar differential inductor. This structure is referred hereafter as multilayer conventional symmetric inductor. Realization of cost-effective symmetric inductor structures with minimum area without performance degradation is addressed in this chapter. This chapter presents the design of a different form of multilevel symmetric winding in which the traces of the metal spiral up and down in a pyramidal manner and hence called as multilayer pyramidal symmetric (MPS) inductor. The design of the MPS inductor is discussed in Sect. 3.2. By varying the width, diameter and metal trace offsets between the adjacent metal layers, the performance of the MPS inductors are evaluated in Sect. 3.5. A compact

G. Haobijam and R. P. Palathinkal, *Design and Analysis of Spiral Inductors*,
DOI: 10.1007/978-81-322-1515-8_3,

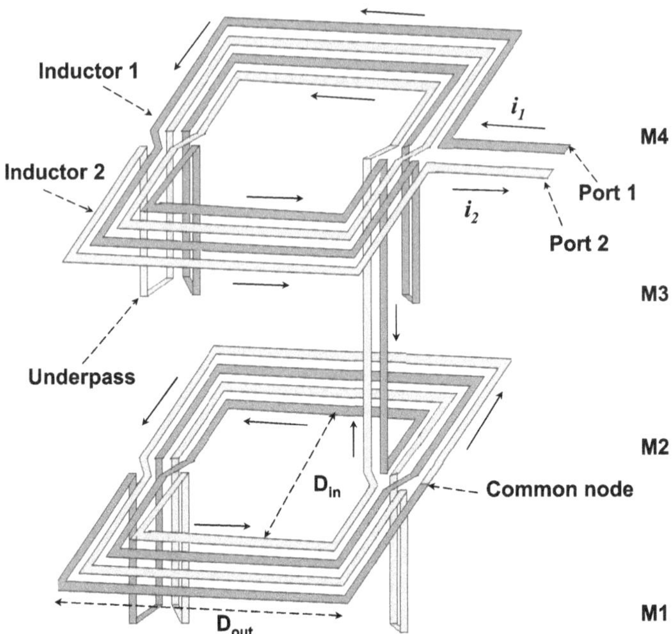

Fig. 3.1 Layout of a two layer conventional symmetric inductor using four metal layers (not drawn to scale)

model to predict the equivalent parasitic capacitance and self resonant frequency of the structure is also developed in Sect. 3.3. The structure is fabricated and the testing and the measured results are discussed in Sect. 3.6 and finally the design, fabrication, and testing of MPS inductor is summarized in Sect. 3.7. It may be noted that in this Chapter, five different inductor structures are referred repeatedly. Three of them are multilayer structures viz. MPS, multilayer conventional symmetric structure as shown in Fig. 3.1 and multilayer conventional asymmetric stack. The other two structures are conventional planar symmetric as shown in Fig. 1.10 and the symmetrical inductor using a pair of planar asymmetrical inductor as shown in Fig. 1.9.

3.2 Design of MPS Inductor Structure

The multilayer pyramidal symmetric inductor (MPS) is shown in Fig. 3.2. The MPS inductor is realized by connecting two inductors in series as indicated by Inductor 1 and Inductor 2 in the figure. In these inductors, the metal traces are pyramidically wound and are referred as pyramidal inductor. For example, for a four layer MPS structure, Inductor 1 has its outermost (first) turn on topmost metal, M4 and the second turn on M3 and follows similarly down to the bottom metal level,

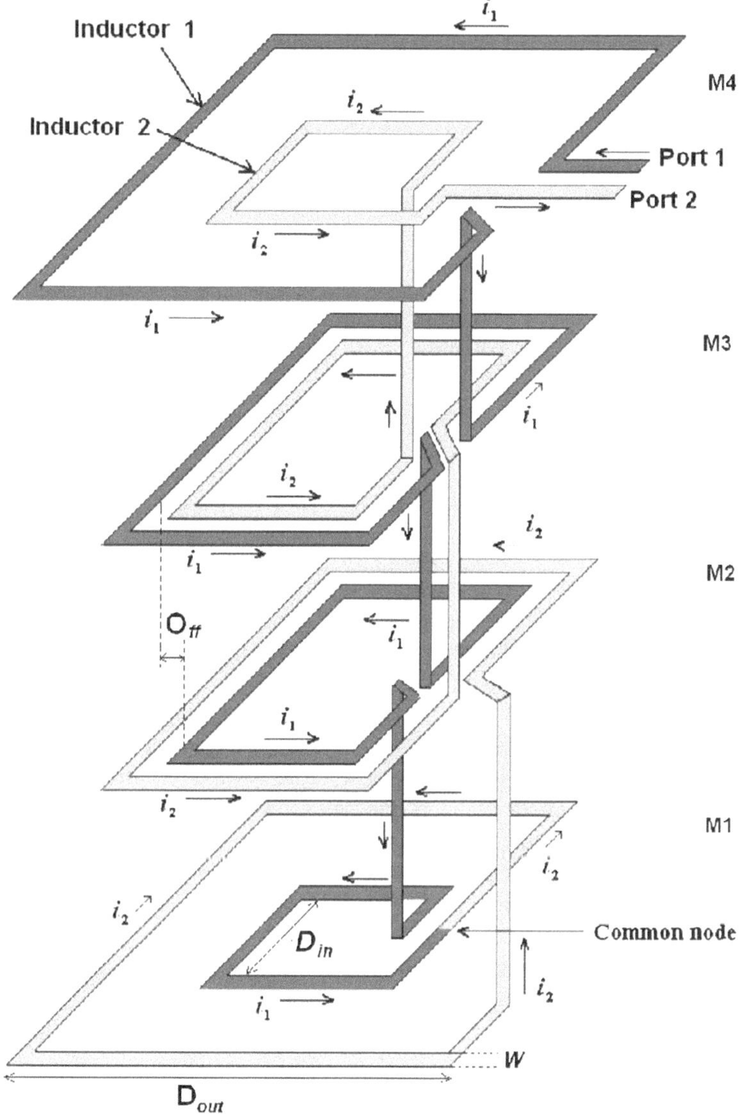

Fig. 3.2 Multilayer pyramidal symmetric inductor

M1 having the innermost (fourth) turn. The second pyramidal spiral inductor starts winding from bottom metal level, M1 with outermost (first) turn and the second turn on M2 and repeats till it ends with the innermost turn on M4. The two inductors are connected at the bottom metal level, M1. So for each inductor, the turns of the metal track runs in different metal layers and the inner diameter changes proportionately to avoid the overlapping of the turns and spirals down and up in a pyramidal manner and hence the name. There will be coupling between these two inductors and the total inductance is given by the sum of its self and mutual inductances. If W is the metal width and O_{ff} is the offset between the edges of two consecutive turns in adjacent metal layer, then the inner diameter decreases or increases by $2(W + O_{ff})$ as the inductor winds down and up respectively. The jth turn of one inductor overlaps with the same jth turn of the other inductor as seen in Fig. 3.2 and winding up and down is achieved using two vias between adjacent metal layers. The design parameters comprise of W, O_{ff}, D_{out}, D_{in} and number of metal layers. Since each inductor has only one turn in each layer, there are only two metal traces in each layer. The effective increase in series resistance due to current crowding is expected to be small as compared to multilayer conventional symmetric structure. The upper metal layers of the process technology is used to avoid the thinner lower metal layers and associated substrate losses.

3.3 Lumped Element Model of the MPS Inductor Structure

In this section, a lumped element model of the multilayer pyramidal symmetric inductor is developed based on the models developed for single layer and multilayer structures [6–8]. The model is shown only for a four layer MPS inductor in Fig. 3.3. As discussed in the previous section, the MPS inductor can be considered as two inductors connected in series. So, the symmetric inductor designed using four metal layers can be modeled as two inductors of four turns each, connected in series. Considering each turn as a segment, the structure is divided into eight segments. The length of each segment is represented by l_{ij} where i denotes inductor 1 or 2 and j denotes the turn number 1, 2, 3 or 4 of each inductor. However, the length of the jth turn of both the inductors are equal, i.e., $l_{11} = l_{21}$, etc. The inductance of the jth turn of the ith inductor is also represented by L_{ij}. For example L_{12} represents inductance of second turn of the first inductor. The jth turn of e ach inductor overlaps with each other and the overlapping capacitances are given by C_{jj}. The coupling to the substrate will be only from L_{21}, L_{22}, L_{13}, and L_{14} as it can be easily observed in Fig. 3.2. That means the coupling from L_{11}, L_{12}, L_{23}, and L_{24} will be shielded by L_{21}, L_{22}, L_{13}, and L_{14}, respectively. The respective oxide capacitances are represented by C_{ox_j}. The C_{sub} and R_{sub} denote the substrate parasitics. Therefore, there will be a total of four metal to metal overlap capacitances and four metal to substrate capacitances. The metal trace to trace capacitance is usually smaller than the metal to metal overlap capacitance and hence neglected here for simplicity [9]. Following a similar approach, the model can be extended for N layers where N is even. The number of metal layers used for designing the inductor is equal to the number of turns. Hence for an MPS inductor designed using N layer, the structure can be divided into 2N segments and there will be N metal to metal overlap capacitance and N metal to substrate capacitance.

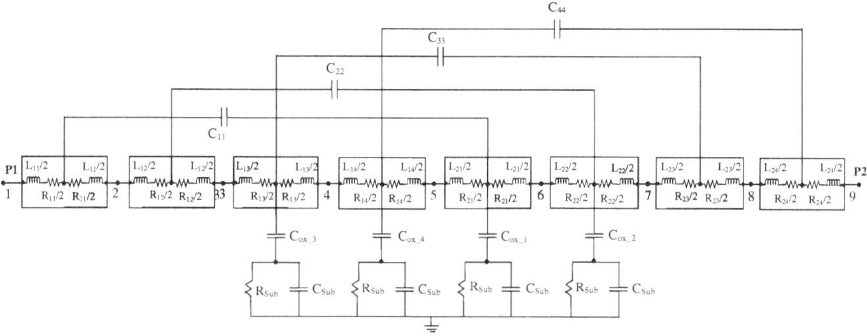

Fig. 3.3 Lumped element model of a four layer multilayer pyramidal symmetric inductor

The self resonant frequency of an inductor is defined as the frequency at which the peak magnetic energy becomes equal to the electric energy, i.e., the inductive reactance and the capacitive reactance become equal and opposite. It is given by

$$f_{\text{res}} = \frac{1}{2\pi\sqrt{L_{\text{eqv}} C_{\text{eqv}}}} \tag{3.1}$$

where L_{eqv} is the equivalent inductance and C_{eqv} is the equivalent parasitic capacitance. The C_{eqv} for a given voltage can be estimated from the total electrical energy stored in the structure as expressed by $1/2C_{\text{eqv}}V^2$. The total energy is the energy stored in the equivalent metal to metal (E_{mm}) and metal to substrate capacitance (E_{mSub}),

$$E_{\text{total}} = E_{\text{mm}} + E_{\text{mSub}} \tag{3.2}$$

The voltage profiles along with the distributed capacitances of the structure is shown in Fig. 3.4. Assuming a linear voltage profile [9, 10] the voltage at each node m is given by

$$V_m = \begin{cases} \frac{V_o}{2} - \frac{V_o}{2}\frac{\sum_{j=1}^{m-1} l_{1j}}{\sum_{j=1}^{4} l_{1j}} & 1 \le m \le 4 \\ 0 & m = 5 \\ \frac{-V_o}{2} - \frac{-V_o}{2}\frac{\sum_{j=m-4}^{4} l_{2j}}{\sum_{j=1}^{4} l_{2j}} & 6 \le m \le 9 \end{cases} \tag{3.3}$$

The voltage across each turn or segment of the inductor is denoted by V_{ij}, where i is the inductor 1 or 2 and j is the turn number 1, 2, 3, or 4 of each inductor. V_{ij} can be calculated by averaging the voltage at its two nodes. For example

$$V_{11} = \frac{1}{2}(V_1 + V_2) \tag{3.4}$$

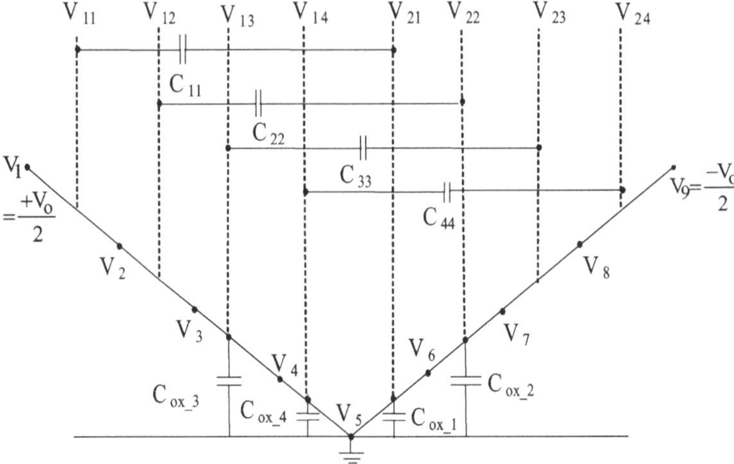

Fig. 3.4 Voltage profile and distributed capacitance of multilayer pyramidal symmetric inductor

In order to compute the energy stored at each C_{jj}, the voltage drop between the jth turn of each inductor is given by

$$\Delta V_{1j,2j} = V_{1j} - V_{2j} \tag{3.5}$$

It can be seen that

$$\Delta V_{11,21} = \Delta V_{12,22} = \Delta V_{13,23} = \Delta V_{14,24} = \frac{V_o}{2} \tag{3.6}$$

Then, the energy stored in the equivalent metal to metal capacitance (E_{mm}) and metal to substrate capacitance (E_{mSub}) of a four layer MPS inductor can be calculated as follows.

$$\begin{aligned}
E_{\text{mm}} &= \frac{1}{2}\, C_{\text{eqv_mm}} {V_o}^2 \\
&= \frac{1}{2}\, C_{11} \Delta V_{11,21}^2 + \frac{1}{2}\, C_{22} \Delta V_{12,22}^2 + \frac{1}{2}\, C_{33} \Delta V_{13,23}^2 + \frac{1}{2}\, C_{44} \Delta V_{14,24}^2 \\
&= \frac{1}{2}\, C_{M_1 M_4} A_1 \Delta V_{11,21}^2 + \frac{1}{2}\, C_{M_2 M_3} A_2 \Delta V_{12,22}^2 + \frac{1}{2}\, C_{M_2 M_3} A_3 \Delta V_{13,23}^2 \\
&\quad + \frac{1}{2}\, C_{M_1 M_4} A_4 \Delta V_{14,24}^2 \\
&= \frac{1}{2}\, V_o^2\, \frac{C_{M_1 M_4}(A_1 + A_4) + C_{M_2 M_3}(A_2 + A_3)}{4}
\end{aligned} \tag{3.7}$$

$$
\begin{aligned}
E_{\mathrm{mSub}} &= \frac{1}{2}\, C_{\mathrm{eqv\_mSub}} V_o{}^2 \\
&= \frac{1}{2}\, C_{\mathrm{ox\_1}} V_{21}^2 + \frac{1}{2}\, C_{\mathrm{ox\_2}} V_{22}^2 + \frac{1}{2}\, C_{\mathrm{ox\_3}} V_{13}^2 + \frac{1}{2}\, C_{\mathrm{ox\_4}} V_{14}^2 \\
&= \frac{1}{2}\, C_{M_1\mathrm{Sub}} A_1 V_{21}^2 + \frac{1}{2}\, C_{M_2\mathrm{Sub}} A_2 V_{22}^2 + \frac{1}{2}\, C_{M_2\mathrm{Sub}} A_3 V_{13}^2 \\
&\quad + \frac{1}{2}\, C_{M_1\mathrm{Sub}} A_4 V_{14}^2 \\
&= \frac{1}{2}\, V_o^2 \frac{1}{16(l_{21} + l_{22} + l_{13} + l_{14})^2} \\
&\quad \times \Big[C_{M_1\mathrm{Sub}} (A_1 l_{21}^2 + A_4 l_{14}^2) \\
&\qquad + C_{M_2\mathrm{Sub}} \Big(A_2 (2l_{21} + l_{22})^2 + A_3 (l_{13} + 2l_{14})^2 \Big) \Big]
\end{aligned} \tag{3.8}
$$

where $C_{M_1M_4}$ and $C_{M_2M_3}$ are the capacitance per unit area between the metal layers 1 and 4 and between 2 and 3, respectively. The $C_{M_1\mathrm{Sub}}$ and $C_{M_2\mathrm{Sub}}$ are the capacitance per unit area of substrate with respect to the metal layer 1 and 2, respectively. A_j is the area of the jth turn in terms of outer diameter D_{out}, metal width W and offset O_{ff} given by

$$
A_j = 4W\big[D_{\mathrm{out}} - W(2j - 1) - 2O_{\mathrm{ff}}(j - 1)\big], \qquad j = 1, 2 \ldots n \tag{3.9}
$$

Therefore, the equivalent metal to metal, $C_{\mathrm{eqv\_mm}}$ and metal to substrate capacitance, $C_{\mathrm{eqv\_mSub}}$ can be calculated as

$$
C_{\mathrm{eqv\_mm}} = \frac{1}{4}\Big[C_{M_1M_4}(A_1 + A_4) + C_{M_2M_3}(A_2 + A_3) \Big] \tag{3.10}
$$

$$
\begin{aligned}
C_{\mathrm{eqv\_mSub}} &= \frac{1}{16(l_{21} + l_{22} + l_{13} + l_{14})^2} \\
&\quad \times \Big[C_{M_1\mathrm{Sub}} (A_1 l_{21}^2 + A_4 l_{14}^2) \\
&\qquad + C_{M_2\mathrm{Sub}} \Big(A_2 (2l_{21}^2 + l_{22})^2 + A_3 (l_{13} + 2l_{14})^2 \Big) \Big]
\end{aligned} \tag{3.11}
$$

Finally, the equivalent capacitance of an N layer MPS inductor is thus given by Eq. 3.12 where $l_{\mathrm{total}} = \sum_{j=1}^{n} l_j$ and $j = 1$ to n. Since all $l_{1j} = l_{2j}$, inductor number is removed from the notation. Similar expressions for a two layer stack of n turns is available in literature [9, 10] and reproduced here in Eq. 3.13, following the same nomenclature of the above derivations. The equivalent capacitance of a two layer conventional symmetric inductor of n turns is also derived following the same approach and is discussed in next section. Here, n represents the number of turns of the spiral inductor in each stack layer and the number of turns of each inductor of the two layer symmetric inductor. $C_{M_{\text{top}}M_{\text{bottom}}}$ is the capacitance per unit area between the top spiral metal layer and bottom spiral metal layer while $C_{M_{\text{bottom}}\text{Sub}}$ is between the bottom spiral metal layer and the substrate. The results are discussed in the following section.

$$
\begin{aligned}
C_{\text{eqv\_MPS}} &= \frac{1}{4}\sum_{k=1}^{N/2} C_{M_k M_{N+1-k}}(A_k + A_{N+1-k}) \\
&+ \frac{1}{16\, l_{\text{total}}^2}\sum_{k=1}^{N/2} C_{M_k\text{Sub}}\left[A_k\left\{\sum_{j=1}^{k} 2l_j - l_k\right\}^2\right. \\
&\left.+ A_{N+1-k}\left\{\sum_{j=1}^{k} 2l_{N+1-j} - l_{N+1-k}\right\}^2\right]
\end{aligned} \tag{3.12}
$$

$$
\begin{aligned}
C_{\text{eqv\_Stack}} &= \frac{1}{4\, l_{\text{total}}^2} C_{M_{\text{top}}M_{\text{bottom}}}\sum_{k=1}^{n} A_k\left[2\sum_{j=k}^{n} l_j - l_k\right]^2 \\
&+ \frac{1}{16\, l_{\text{total}}^2} C_{M_{\text{bottom}}\text{Sub}}\sum_{k=1}^{n} A_k\left[2\sum_{j=k}^{n} l_j - l_k\right]^2
\end{aligned} \tag{3.13}
$$

3.4 Parasitic Capacitance Calculation for Conventional Multilayer Symmetric Inductor

Let us first consider the two layer structure as in Fig. 3.1 where each inductor has four turns. For each inductor, a half turn of the spiral can be considered as a segment and the structure can be broken down into eight segments. Let the lengths of each segment of the inductor be represented by S_{ij} where i denotes the inductor 1 or 2 and j denotes the segment number 1, 2, 3, 4, 5, 6, 7, and 8 of each inductor. If each end of the segment represent a node then, there will be a total of 17 nodes. The voltage profiles along with the distributed capacitances of the structure is shown in Fig. 3.5. The voltage at each node m is given by

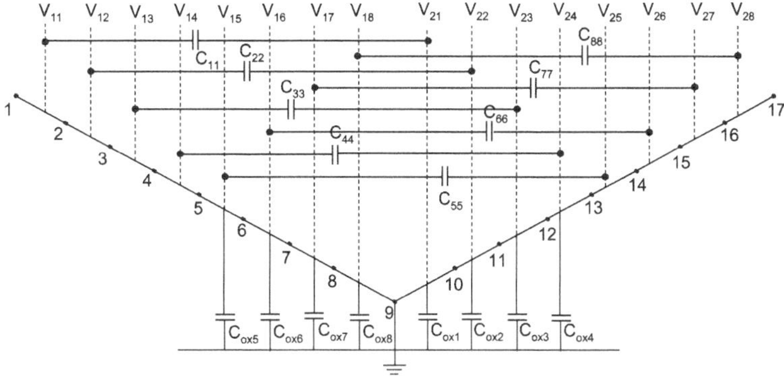

Fig. 3.5 Voltage profile and distributed capacitance of two layer conventional symmetric inductor

$$V_m = \begin{cases} \frac{V_o}{2} - \frac{V_o}{2} \frac{\sum_{j=1}^{m-1} S_{1j}}{\sum_{j=1}^{8} S_{1j}} & 1 \le m \le 8 \\ 0 & m = 9 \\ \frac{-V_o}{2} - \frac{-V_o}{2} \frac{\sum_{j=m-8}^{8} S_{2j}}{\sum_{j=1}^{8} S_{2j}} & 10 \le m \le 17 \end{cases}$$

The voltage across each segment of the inductor denoted by V_{ij} where i is the inductor 1 or 2 and j is the segment number 1, 2, 3,..., 8 of each inductor can be calculated as

$$V_{ij} = \frac{1}{2}(V_j + V_{j+1}) \quad \text{for } i = 1 \text{ to } 2,\ j = 1 \text{ to } 8 \tag{3.14}$$

The energy stored at each metal to metal overlapping capacitance C_{jj} in Fig. 3.5 can be calculated by calculating the voltage drop between the jth segment of each ith inductor as

$$\Delta V_{1j,2j} = V_{1j} - V_{2j} \tag{3.15}$$

From this, it will result that $\Delta V_{11,21} = \Delta V_{12,22} = \Delta V_{13,23} = \Delta V_{14,24} = \Delta V_{15,25} = \Delta V_{16,26} = \Delta V_{17,27} = \Delta V_{18,28} = \frac{V_o}{2}$. Then, the energy stored in the equivalent metal to metal capacitance (E_{mm}) and metal to substrate capacitance (E_{mSub}) of a four layer MCS inductor can be calculated as follows.

$$
\begin{aligned}
E_{\mathrm{mm}} &= \frac{1}{2} C_{\mathrm{eqv\_mm}} V_o{}^2 \\
&= \frac{1}{2} C_{11} \Delta V_{11,21}^2 + \frac{1}{2} C_{22} \Delta V_{12,22}^2 + \frac{1}{2} C_{33} \Delta V_{13,23}^2 \\
&\quad + \frac{1}{2} C_{44} \Delta V_{14,24}^2 + \frac{1}{2} C_{55} \Delta V_{15,25}^2 + \frac{1}{2} C_{66} \Delta V_{16,26}^2 \\
&\quad + \frac{1}{2} C_{77} \Delta V_{17,27}^2 + \frac{1}{2} C_{88} \Delta V_{18,28}^2 \\
&= \frac{1}{2} C_{M_2M_4} a_1 \Delta V_{11,21}^2 + \frac{1}{2} C_{M_2M_4} a_2 \Delta V_{12,22}^2 \\
&\quad + \frac{1}{2} C_{M_2M_4} a_3 \Delta V_{13,23}^2 + \frac{1}{2} C_{M_2M_4} a_4 \Delta V_{14,24}^2 \\
&\quad + \frac{1}{2} C_{M_2M_4} a_5 \Delta V_{15,25}^2 + \frac{1}{2} C_{M_2M_4} a_6 \Delta V_{16,26}^2 \\
&\quad + \frac{1}{2} C_{M_2M_4} a_7 \Delta V_{17,27}^2 + \frac{1}{2} C_{M_2M_4} a_8 \Delta V_{18,28}^2 \\
&= \frac{1}{2} V_o^2 \, C_{M_2M_4} \frac{a_1 + a_2 + a_3 + a_4 + a_5 + a_6 + a_7 + a_8}{4} \\
&= \frac{1}{2} V_o^2 \, C_{M_2M_4} \frac{A_1 + A_2 + A_3 + A_4}{4}
\end{aligned}
\tag{3.16}
$$

where A_1, A_2, A_3, and A_4 represent the total area of the spiral turn 1, 2, 3, and 4. For an easy comparison and to comply with the notations of the stack and the MPS inductors, this representation is adopted since $A_1 \cong (a_1 + a_8)$, $A_2 \cong (a_2 + a_7)$, $A_3 \cong (a_3 + a_6)$, and $A_4 \cong (a_4 + a_5)$.

$$
\begin{aligned}
E_{\mathrm{mSub}} &= \frac{1}{2} C_{\mathrm{eqv\_mSub}} V_o{}^2 \\
&= \frac{1}{2} C_{\mathrm{ox\_1}} V_{21}^2 + \frac{1}{2} C_{\mathrm{ox\_2}} V_{22}^2 + \frac{1}{2} C_{\mathrm{ox\_3}} V_{23}^2 + \frac{1}{2} C_{\mathrm{ox\_4}} V_{24}^2 \\
&\quad + \frac{1}{2} C_{\mathrm{ox\_5}} V_{15}^2 + \frac{1}{2} C_{\mathrm{ox\_6}} V_{16}^2 + \frac{1}{2} C_{\mathrm{ox\_7}} V_{17}^2 + \frac{1}{2} C_{\mathrm{ox\_8}} V_{18}^2 \\
&= \frac{1}{2} C_{M_2\mathrm{Sub}} a_1 V_{21}^2 + \frac{1}{2} C_{M_2\mathrm{Sub}} a_2 V_{22}^2 + \frac{1}{2} C_{M_2\mathrm{Sub}} a_3 V_{23}^2 \\
&\quad + \frac{1}{2} C_{M_2\mathrm{Sub}} a_4 V_{24}^2 + \frac{1}{2} C_{M_2\mathrm{Sub}} a_5 V_{15}^2 + \frac{1}{2} C_{M_2\mathrm{Sub}} a_6 V_{16}^2 \\
&\quad + \frac{1}{2} C_{M_2\mathrm{Sub}} a_7 V_{17}^2 + \frac{1}{2} C_{M_2\mathrm{Sub}} a_8 V_{18}^2
\end{aligned}
\tag{3.17}
$$

On calculation of voltages from Eq. 3.15, we will get $V_{18} = -V_{21}$, $V_{17} = -V_{22}$, $V_{16} = -V_{23}$, and $V_{15} = -V_{24}$. Also, we can see that $S_{18} = S_{21} = \frac{l_1}{2}$, $S_{17} = S_{22} = \frac{l_2}{2}$, $S_{16} = S_{23} = \frac{l_3}{2}$ and $S_{15} = S_{24} == \frac{l_4}{2}$, where l_1, l_2, l_3, and l_4 are the total length of the full turn 1, 2, 3, and 4. Turn 1 is the outermost one. Therefore, the above equation further simplifies to

$$
\begin{aligned}
E_{\text{mSub}} &= \frac{1}{2} V_o^2 \, C_{M_2\text{Sub}} \frac{1}{16\, l_{\text{total}}^2} \Big[S_{18}^2 (a_1 + a_8) \\
&\quad + (2S_{18} + S_{17})^2 \, (a_2 + a_7) + (2S_{18} + 2S_{17} + S_{16})^2 (a_3 + a_6) \\
&\quad + (2S_{18} + 2S_{17} + 2S_{16} + S_{15})^2 (a_4 + a_5) \Big] \\
&= \frac{1}{2} V_o^2 \, C_{M_2\text{Sub}} \frac{1}{16\, l_{\text{total}}^2} \left[A_1 \left(\frac{l_1}{2}\right)^2 + A_2 \left(\frac{2l_1 + l_2}{2}\right)^2 \right. \\
&\quad \left. + A_3 \left(\frac{2l_1 + 2l_2 + l_3}{2}\right)^2 + A_4 \left(\frac{2l_1 + 2l_2 + 2l_3 + l_4}{2}\right)^2 \right]
\end{aligned}
\tag{3.18}
$$

$C_{M_2M_4}$ is the capacitance per unit area between the metal layer 4, i.e., the metal layer of the top spiral and 2, i.e., the metal layer of the bottom spiral. $C_{M_2\text{Sub}}$ is the capacitance per unit area between substrate and the bottom spiral metal layer 2. Therefore, the equivalent metal to metal, $C_{\text{eqv\_mm\_MCS}}$ and metal to substrate capacitance, $C_{\text{eqv\_mSub\_MCS}}$ will be given by

$$
C_{\text{eqv\_mm\_MCS}} = \frac{1}{4} C_{M_2M_4} \, (A_1 + A_2 + A_3 + A_4) \tag{3.19}
$$

$$
\begin{aligned}
C_{\text{eqv\_mSub\_MCS}} = & \frac{1}{16\, l_{\text{total}}^2} \, C_{M_2\text{Sub}} \left[A_1 \left(\frac{l_1}{2}\right)^2 \right. \\
& + A_2 \left(\frac{2l_1 + l_2}{2}\right)^2 + A_3 \left(\frac{2l_1 + 2l_2 + l_3}{2}\right)^2 \\
& \left. + A_4 \left(\frac{2l_1 + 2l_2 + 2l_3 + l_4}{2}\right)^2 \right]
\end{aligned}
\tag{3.20}
$$

The area of the full spiral jth turn can be calculated in terms of outer diameter D_{out}, metal width W, and turn to turn spacing s given by

$$
A_j = 4W \big[D_{\text{out}} - W(2j - 1) - 2s(j - 1) \big], \quad j = 1, 2 \ldots\ldots n \tag{3.21}
$$

Thus for a two layer conventional symmetric inductor of n turns, the equivalent parasitic capacitance $C_{\text{eqv\_MCS}}$ is given by Eq. 3.22.

$$C_{\text{eqv\_MCS}} = \frac{1}{4} C_{M_{\text{top}} M_{\text{bottom}}} \sum_{k=1}^{n} A_k + \frac{1}{16\, l_{\text{total}}^2} C_{M_{\text{bottom}} \text{Sub}} \sum_{k=1}^{n} A_k \left[\frac{l_k}{2} + 2 \sum_{j=2}^{k} \frac{l_{j-1}}{2} \right]^2 \tag{3.22}$$

3.5 Characterization of the MPS Inductor Structure

In this section, the performance of several MPS inductors of varying width, diameter, and metal offsets are discussed. Subsequently, the performance and area is compared with the asymmetric pair and single layer (planar) symmetric structure. The performance is also compared to other single layer symmetrical inductors reported in the literature. Third, the equivalent parasitic capacitance and self resonating frequency of MPS, two layer asymmetric stack and two layer conventional symmetric, each having the same layout parameters, are compared using the analytical expressions of the previous section. This predictions are also validated with EM simulation results in each case. The results of inductance and the quality factor presented in this section are calculated from the two port parameters of EM simulation results, both for single ended and differential excitation. The single ended and differential impedance represented by Z_{se} and Z_{diff} is calculated as

$$Z_{\text{se}} = \frac{1}{Y_{11}} \tag{3.23}$$

$$Z_{\text{diff}} = \frac{Y_{11} + Y_{22} + Y_{12} + Y_{21}}{Y_{11}Y_{22} - Y_{12}Y_{21}} \tag{3.24}$$

where Y_{11}, Y_{12}, Y_{21}, Y_{22} are the admittance or Y parameters. The inductance and quality factor for single ended and differential configuration denoted by L_{se}, Q_{se}, L_{diff} and Q_{diff} respectively is calculated as

$$L_{\text{se}} = \frac{Im(Z_{\text{se}})}{2\pi f} \tag{3.25}$$

$$Q_{\text{se}} = \frac{Im(Z_{\text{se}})}{Re(Z_{\text{se}})} \tag{3.26}$$

$$L_{\text{diff}} = \frac{Im(Z_{\text{diff}})}{2\pi f} \tag{3.27}$$

$$Q_{\text{diff}} = \frac{Im(Z_{\text{diff}})}{Re(Z_{\text{diff}})} \tag{3.28}$$

Table 3.1 Layout parameters and figure of merits of MPS inductors.

Group	Width (μm)	Offsets (μm)	D_{in} (μm)	D_{out} (μm)	L (nH)	Q_{max}	f_{max} (GHz)	f_{res} (GHz)	L/A (pH/μm^2)
A	8	2	174	250	21.4	4.0	1.0	2.0	0.34
	8	2	130	206	14.2	4.7	1.2	2.6	0.33
	8	2	54	130	8	6.1	2.4	6.2	0.47
B	8	2	146	222	16.8	4.5	1.1	2.4	0.34
	12	2	114	222	14.0	5.1	1.0	2.5	0.28
	16	2	82	222	10.8	5.3	1.0	2.7	0.21
C	12	2	114	222	14.0	5.1	1.0	2.5	0.28
	12	4	102	222	12.4	5.0	1.1	2.6	0.25
	12	6	90	222	10.8	5.1	1.2	2.9	0.21

3.5.1 Performance Trend of MPS Inductors

The figure of merit (FOM) of on-chip spiral inductors are (i) quality factor, Q (ii) optimum frequency, f_{max} at which Q reaches its maximum value, Q_{max} (iii) self-resonant frequency, f_{res} at which the inductor behaves like a parallel RC circuit in resonance and is far from behaving as an inductor [11], and (iv) inductance to silicon area ratio (L/A). To investigate the effect of width, diameter, and metal offsets on the figure of merits, three groups of inductors were simulated, wherein one of the parameter is varied keeping the other two constant. The layout parameters of these inductors are given in Table 3.1. In Group A the diameter is varied, in Group B the width is varied, and in Group C the metal offset is varied.

The structures are simulated in a six metal layer 0.18 μm process technology using a 3D Electromagnetic simulator[1] [12]. In simulation, the substrate and the dielectric layers are defined as per the technology parameters of the process to reproduce the actual inductor as close as possible. The performance trend is demonstrated here for single ended applications and therefore, the inductances and the quality factors are calculated according to Eqs. 3.25 and 3.26, respectively. In Group A MPS inductors, as the outer diameter decreases the total length of the metal trace will decrease while other parameters are kept constant. As a result the inductance decreases and quality factor increases as shown in Fig. 3.6a, b, respectively. In Group B and C, the outer diameter is kept constant and the width and metal offset between the adjacent metal layer is varied. The inner diameter decreases with the increase in width and offset. This will shorten the total length of the spiral and therefore the inductance value decreases. The quality factor will thus increase with a decrease in inductance. The variation in inductance and quality factor is shown in Figs. 3.7 and 3.8. The self resonating frequency is higher for smaller inductance in all the cases. The inductance, peak quality factor, and the self resonant frequency are also given in Table 3.1. The variation of the diameter in Group A results in a significant change in the inductance

[1] Intellisuite, Intellisense Software Corp.

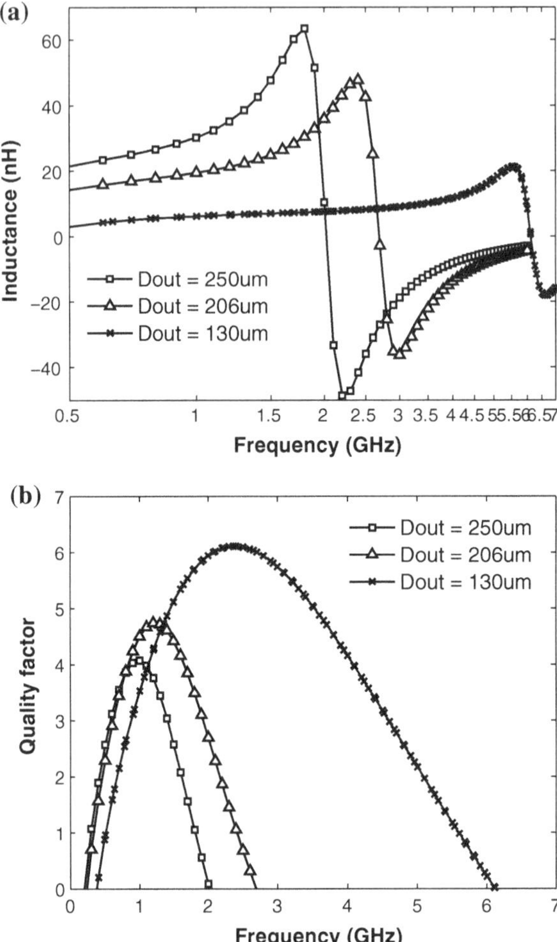

Fig. 3.6 **a** Inductance and **b** Quality factor of Group A MPS inductors with different outer diameters. Width and offset is kept constant at 8 and 2 μm, respectively

and quality factor in contrast to the variation in metal width and offset. The metal offset is analogs to the inter turn spacing in planar structures. In Group C, as the metal offset increases, the inductance decreases whereas the quality factor is almost constant. This shows that with a small offset the magnetic coupling can be maximized and the inductance to area ratio can be increased. These results are also consistent with the performance trend of planar inductors with layout parameters studied in [13–15]. The important characteristic of MPS structure is its symmetrical nature. In Fig. 3.9, the input impedance seen at port 1 and 2 of three different inductors of diameters 250, 222, and 206 μm are plotted. The width and metal offset are 8 and

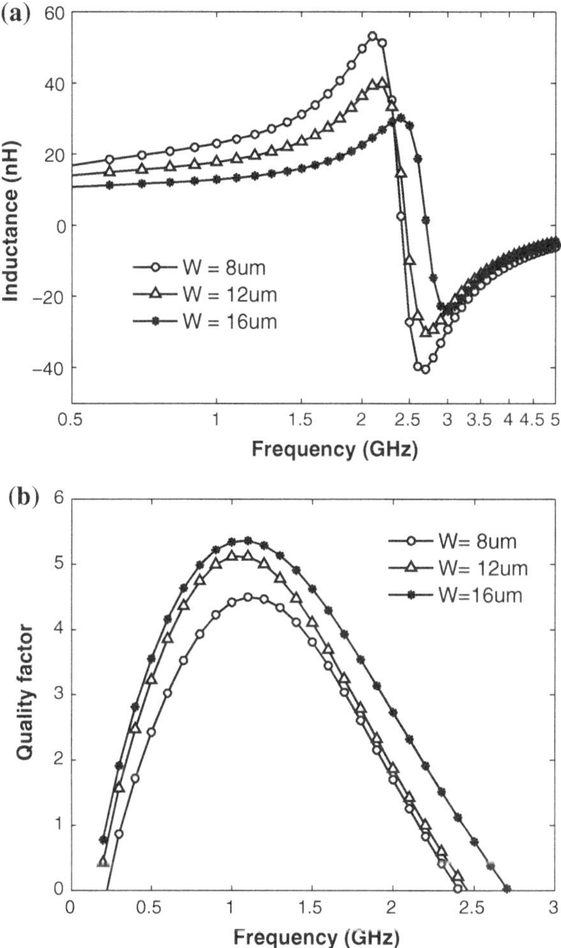

Fig. 3.7 **a** Inductance and **b** Quality factor of MPS inductors (Group B) with different widths of the metal trace. Outer diameter and offset is kept constant at 222 and 2 μm, respectively

2 μm, respectively. The identical impedance measured at each port clearly indicates the symmetry of the structure.

The effect of variation of process parameters on the inductance, quality factor and the self resonance frequency of the MPS inductors was also studied. The results are given in Table 3.2. It would be indeed difficult to present the effect of process parameter variations quantitatively as this would require an extensive simulation of a large number of structures for a large range of inductance values. It may be possible to do a Monte Carlo simulation with the help of some tools, but the MPS is a new structure and since it is not a part of the Foundry design kit, it would be difficult. So only a qualitative result is presented here. We can see that the inductance does

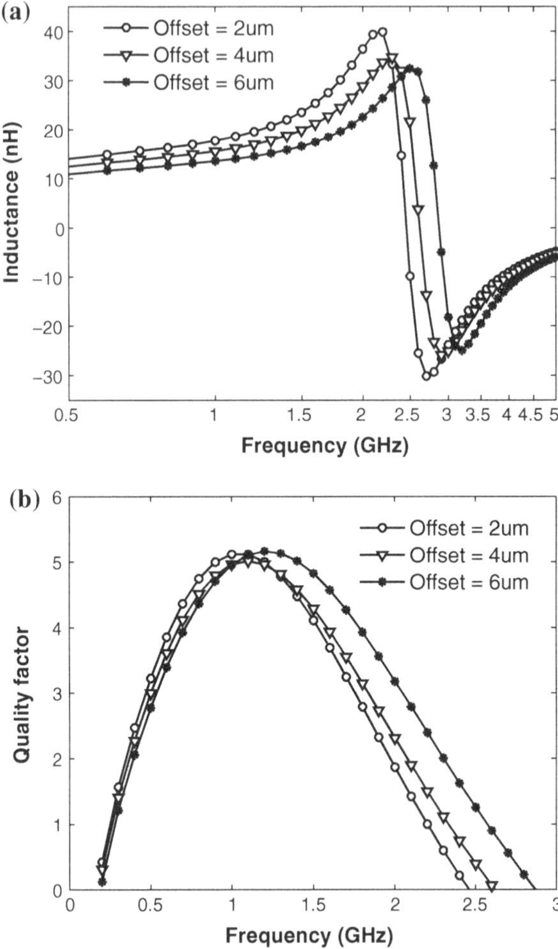

Fig. 3.8 **a** Inductance and **b** Quality factor of MPS inductors (Group C) with different offsets between the adjacent metal layers. Width and outer diameter is kept constant at 12 and 222 μm, respectively

not change with the variation in the process parameters. So the inductance value is determined by its layout parameters. The quality factor increases with the increase in substrate resistivity, oxide thickness and metal thickness and sheet resistance. The self resonance frequency also increases with the increase in substrate resistivity and oxide thickness.

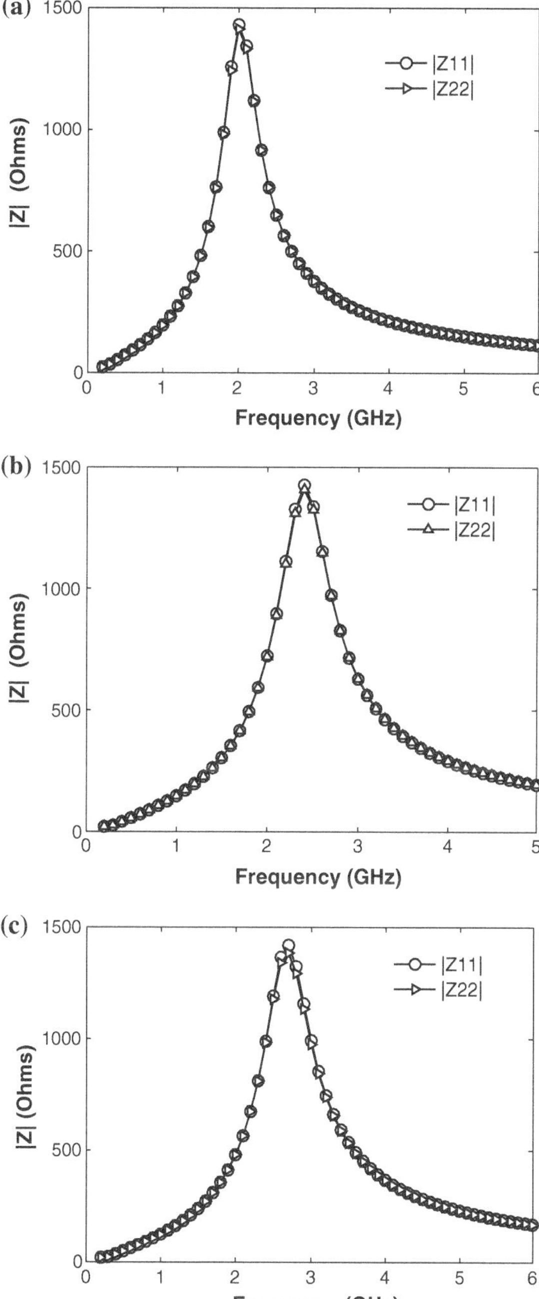

Fig. 3.9 Impedance seen at each port of MPS inductors of **a** Width = 8 μm, D_{out} = 250 μm, Offset = 2 μm (Group A), **b** Width = 8 μm, D_{out} = 222 μm, Offset = 2 μm (Group B) and **c** Width = 8 μm, D_{out} = 206 μm, Offset = 2 μm (Group A)

Table 3.2 Performance trend with variation of process parameters

Process parameters	Trend	Inductance	Quality factor	Self resonance frequency
Metal thickness	Increases	...	Increases	...
Metal sheet resistance	Increases	...	Decreases	...
Substrate resistivity	Increases	...	Increases	Increases
Oxide thickness	Increases	...	Increases	Increases

Note ... negligible change observed

3.5.2 Comparison of MPS with Its Equivalent Planar Inductor Structures

To illustrate the effective area reduction we compared the area required by the proposed MPS structure with its equivalent planar conventional symmetric and a pair of asymmetric inductors. Monolithic inductors are mostly used in the 1–3 GHz frequency range and therefore the comparison is done for an inductance of ≅8 nH at 2 GHz. The area comparison is given in Table 3.3. The width and spacing are kept constant at 8 and 2 μm, respectively. For the case of an asymmetric pair the given area is the approximate area that will be occupied by two 4 nH inductors separated by a distance of 40 μm. The number of turns for each inductor of the asymmetric pair is given.

The last column gives the reduction in area obtained if the same inductance is realized with MPS structure. For the single layer structures we know that the same inductance can be realized with different turns. The layout design parameters of these structures is determined from the layout parameter bounding method discussed in the previous chapter. For any desired inductance value the combination of metal width, number of turns, and the inner and outer diameter is determined. For metal

Table 3.3 Comparison of area occupied by different symmetric inductors structures of 8 nH

Types of inductor structure	Number of turns	Metal layers	Area (μm^2)	% reduction in area of MPS	L/A ($pH/\mu m^2$)
MPS	–	**M6, M5 M4, M3**	**130 × 130**	–	**0.47**
	2	M6	622 × 622	95.6	0.02
Conventional	3	M6	380 × 380	88.2	0.05
planar symmetric	4	M6	286 × 286	79.3	0.09
	5	**M6**	**243 × 243**	**71.3**	**0.16**
	6	M6	220 × 220	65.0	
	2	M6	367 × 774	94.0	0.02
Symmetric inductor	3	M6	238 × 516	86.2	0.05
using a pair of	**4**	**M6**	**191 × 422**	**79.0**	**0.09**
asymmetric 4 nH	5	M6	171 × 382	74.1	0.12
	6	M6	162 × 364	71.3	0.13

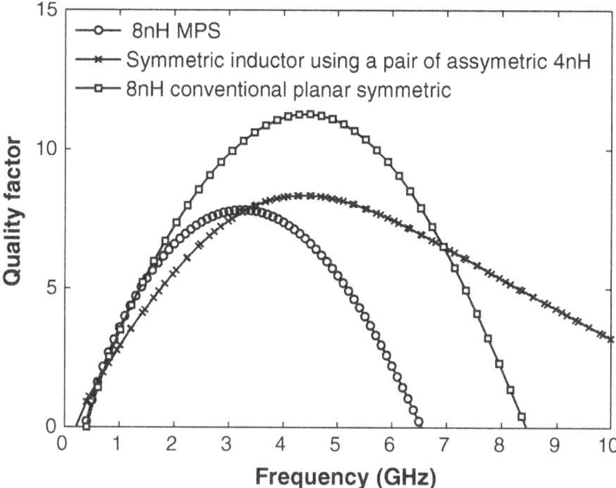

Fig. 3.10 Quality factor comparison of 8 nH MPS inductors with conventional planar symmetric inductors

width of 8 μm and spacing of 2 μm, the number of turns ranges from 2 to 6 with inner diameter greater than 100 μm. The MPS inductor occupies an area of only $130 \times 130\ \mu m^2$ and hence achieves an area reduction of 65–95 % over its equivalent symmetric inductors, 71–94 % over the pair of asymmetric 4 nH inductors. The inductance to area ratio increases at an average of more than 400 % as compared to planar structures. In the literature, different inductor structures are reported with different technologies. It is difficult to compare the performance very closely. So for comparison the structures are simulated with the same parameters of 0.18 μm technology. The simulated structures are 5 turn conventional symmetric, asymmetric pair of 4 turn each as highlighted in Table 3.3. In order to compare the performance of the MPS structure for differential applications, the quality factor is computed according to Eq. 3.28 and plotted in Fig. 3.10. Both the symmetric and the asymmetric inductor pair is built on M6. At low frequency the quality factor of the MPS inductor and the conventional symmetric inductor are almost the same but higher than the planar asymmetric pair. The planar asymmetric pair has the highest self resonating frequency. The MPS structure, being multilevel has higher parasitics and therefore the resonance frequency is lower. For an application at low frequency the MPS structure will be advantageous with its smaller area. For example at 2.4 GHz the MPS structure achieves an increased quality factor of 11 % and an area reduction of 79 % over its equivalent asymmetric pair. Again, the MPS inductor achieved an area reduction of 71.3 % with a 9 % decrease in the quality factor compared to the conventional symmetric structure. The MPS inductor is compared to the conventional planar symmetric inductors of [2] and [16]. The results are summarized in Table 3.4. The quality factor and the self resonant frequency is more or less comparable. However, the area of MPS inductor is smaller by 72.9 % and 78.4 % compared to symmetrical inductors of [2] and [16] respectively. Therefore, the inductance to area ratio is also higher by more than 300 %.

3.5.3 Comparison of MPS with Multilayer Conventional Symmetric and Asymmetric Stack Structures

In this subsection symmetric structures and asymmetric stack, each having the same layout parameters are compared. The inductor structures are designed in a six metal layer 0.18 μm process technology [17] and the results are summarized in Table 3.5. The outer diameters and metal widths of all the inductors are 130 and 8 μm, respectively. The MPS inductor is designed using M6, M5, M4, and M3 and the turn to turn offset of the MPS inductor is 2 μm. The stack consists of two spiral inductors each of four turns with their topmost spiral on metal layer M6 and their bottom spiral on M5 or M4 or M3. Similarly, the multilayer conventional symmetric structures consist of two differential inductors in M6 and M4 connected in series and their crossover underpasses are in M5 and M3, respectively. Each differential inductor has four turns and a spacing of 2 μm. These three types of inductor structures are more or less similar with the difference only in the form of winding of the metal turns. All the structures are also simulated using EM simulator and the effective inductance measured at 2 GHz is given in the table. The parasitic capacitances are computed using the analytical expressions derived in the previous sections. In a typical six metal layer CMOS technology the metal to metal capacitances viz. $C_{M_6M_5} \cong 34$ aF/μm^2, $C_{M_6M_4} \cong 13$ aF/μm^2, $C_{M_6M_3} \cong 9$ aF/μm^2, $C_{M_5M_4} \cong 36$ aF/μm^2, $C_{M_3\text{Sub}} \cong 12$ aF/μm^2, $C_{M_4\text{Sub}} \cong 8$ aF/μm^2 and $C_{M_5\text{Sub}} \cong 6$ aF/μm^2.

From Table 3.5 we can observe that for the stacked inductor, as the distance of separation between each spiral increases, the f_{res} increases while the inductance changes slightly. As the spirals move away from each other the equivalent metal to metal capacitance decreases and metal to substrate capacitance increases. The overall equivalent parasitic capacitance decreases and therefore the resonance frequency increases. The equivalent parasitic capacitance of the proposed MPS inductor is less than the stack inductor built in M6 and M5 metal layer and more than that of M6 and M4 or M3. The equivalent metal to substrate capacitance of the MPS inductor is much lower than the stack in all the cases. The turn to turn interwinding and underpass capacitances are not considered in the calculation and therefore the predicted values are higher than the simulated ones. For the conventional multilayer symmetric case the spirals are on M6 and M4 and therefore the parasitics are smaller than the MPS. However, due to smaller coupling as a result of larger separation the inductance is much smaller eventhough the layout parameters are same. The inductances and the quality factor of MPS and multilayer conventional symmetric inductors are computed according to Eqs. 3.27 and 3.28. The results are plotted in Fig. 3.11. Since stack is asymmetric, it is not included in the plot for comparison. The results are also reflected in Table 3.5. For the stack inductor, the inductance is calculated based on Eq. 3.25 and the self resonating frequency (simulated) is determined from the quality factor calculated, based on Eq. 3.26. At frequencies ≤3 GHz, the quality factor of MPS is slightly higher. This is because in MPS structure there are only two turns in each layer and the current crowding effect due to the eddy current will be less resulting in smaller ac series resistance and hence higher quality factor. The measured effective

Table 3.4 Comparison of MPS with reported planar symmetric inductors

Inductor structures	Tech.	L (nH)	D_{out} (μm)	Width (μm)	Area (μm^2)	Quality factor Single ended (GHz)	Quality factor Differential ended (GHz)	f_{res} (GHz) Single ended	f_{res} (GHz) Differential ended	L/A (pH/μm^2)
[2]	Three metal	8	250	8	62500	6.6 at 1.6	9.3 at 2.5	6.3	7.1	0.12
[16]	Four metal	8	280	10	78400	≈4–4.5 at <2	≈6.5–7.5 at <3.5	≈4–5	≈5–6	0.10
MPS	Six metal	8	130	8	16900	6.1at 2.4	7.8 at 3.2	6.2	6.4	0.47

Note Structures in [2] and [16] are conventional planar symmetric

Table 3.5 Comparison of MPS inductor with two layer asymmetric stack and two layer conventional symmetric inductor

Structure	Metal layers	N	D_{out} (μm)	Area μm^2	L (nH)	C_{eqv_mm} (fF)	C_{eqv_mSub} (fF)	C_{eqv} (fF)	f_{res}(GHz) Predicted	f_{res}(GHz) Simulated	L/A (pH/μm^2)
Stack	M6, M5	4	130	16900	8.4	132.04	5.82	137.87	4.9	4.6	0.49
(Asymmetric)	M6, M4	4	130	16900	7.5	50.48	7.76	58.25	7.6	6.4	0.44
	M6, M3	4	130	16900	7.2	34.95	11.65	46.60	8.6	7.6	0.42
Multilayer symmetric	M6 M4	4	130	16900	6.7	38.27	1.88	40.15	9.7	6.8	0.39
MPS	M6, M5, M4, M3	–	130	16900	7.3	66.24	2.19	68.43	7.1	6.6	0.43

Note Inductance given is the effective Inductance calculated from EM simulation at 2 GHz. The f_{res} of the stack inductor is given for single ended case

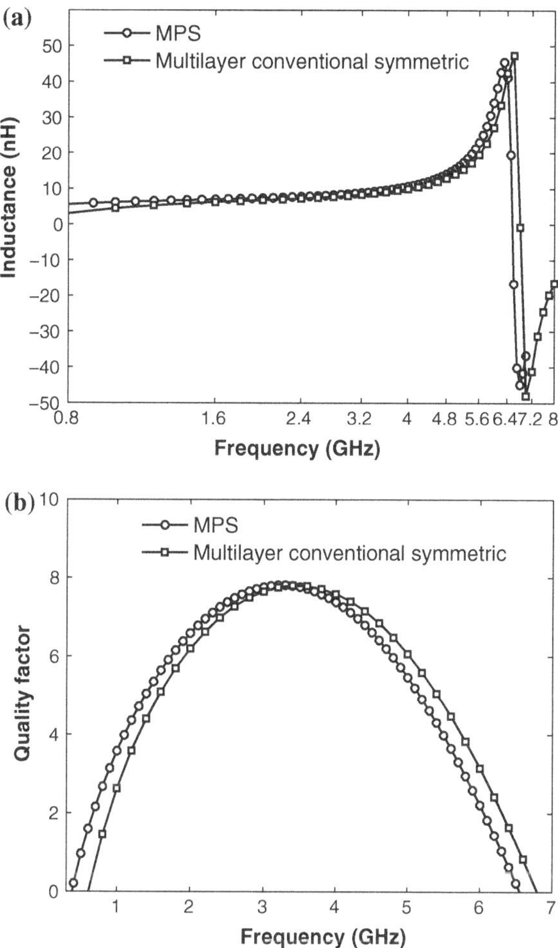

Fig. 3.11 **a** Inductance and **b** Quality factor comparison of MPS and two layer conventional symmetric inductor of same layout parameters

inductances at 2 GHz of MPS and multilayer conventional symmetric are 7.3 nH and 6.6 nH. In fact, to get the same inductance, the outer diameter or the number of turns of the multilayer conventional symmetric must be increased and this will reduce the self resonance frequency and increase the area. So, for multilayer symmetric inductors of equal inductances the MPS structure will have smaller area and higher inductance to area (L/A) ratio with almost comparable performance.

3.6 Experimental Verification

In this section, we discuss the testing and measurement of the MPS inductors. It is presented in three subsections. The process details are briefly stated first and followed by the discussion of the test inductor structures and finally the measured results are reported.

3.6.1 *Process Parameters*

A Few MPS structures were fabricated in 0.18 μm process of United Microelectronics Corporation (UMC). The process parameters are given in Table 3.6. It may be noted that during the design and simulation phase these values are used for the respective parameters.

3.6.2 *Layout of MPS Inductors*

In order to evaluate the performance of the MPS inductors we have fabricated few MPS inductors and the layout parameters are given in Table 3.7. As silicon area is very expensive fabrication of many inductors are not practical. Therefore, we have decided to fabricate two MPS inductors of outer diameter 130 and 222 μm suitable for on-wafer testing. Both the inductors have the same width of 8 μm. The layout parameters were chosen such that the resulting inductance value around 8 and 14 nH. As expected the area required will be more for the 14 nH inductor given that

Table 3.6 Technological parameters [17]

Parameter	Values
Substrate resistivity	20 Ωcm
Silicon dielectric constant	11.9
M6 to substrate separation	8.2 μm
Oxide dielectric constant	4

Table 3.7 Layout parameters and figure of merits of MPS inductors

Test inductor name	W (μm)	Offsets (μm)	D_{in} (μm)	D_{out} (μm)	Simulated L (nH)	Q_{max}	f_{max} (GHz)	f_{res} (GHz)	L/A (pH/μm^2)
W8_Dout130	8	2	54	130	8	6.1	2.4	6.2	0.47
W8_Dout222	8	2	146	222	16.8	4.5	1.1	2.4	0.34

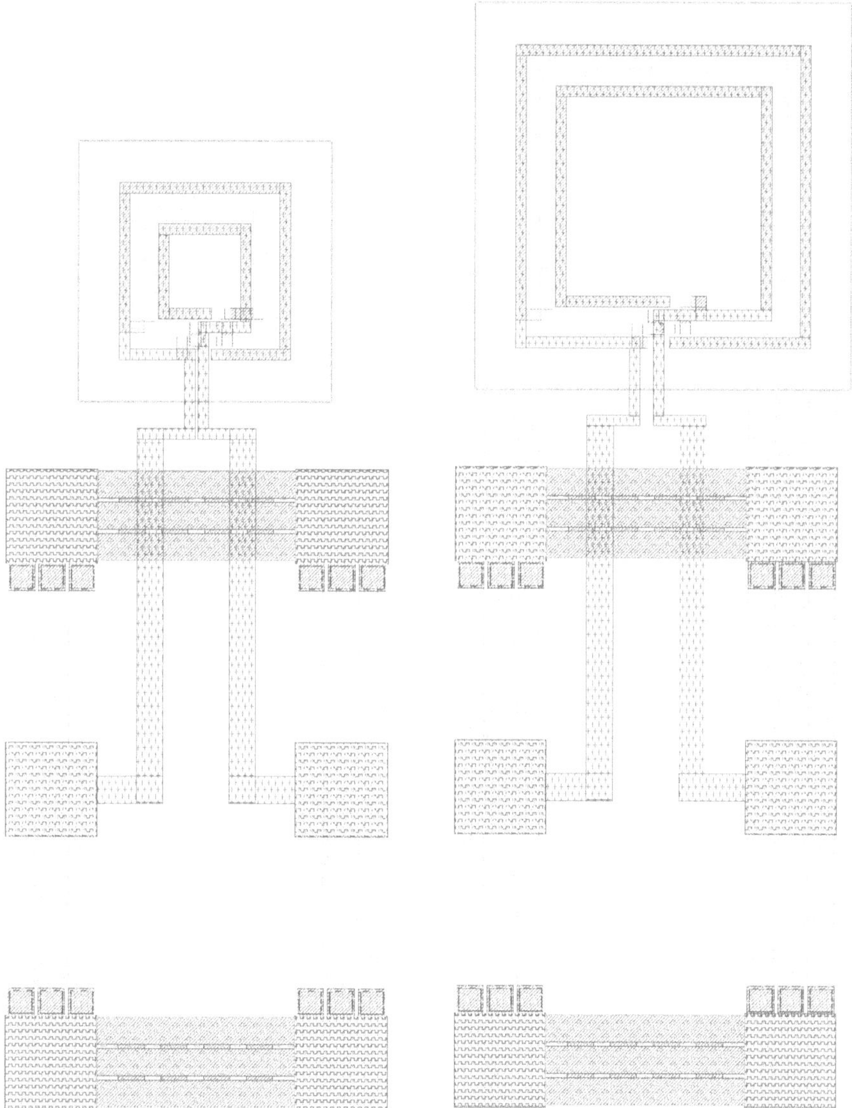

Fig. 3.12 Layout of MPS test inductors W8_Dout130 (*left*) and W8_Dout222 (*right*) in GSG configuration for on-wafer measurement

the offset and the width are the same. The inductor of outer diameter 130 μm will be used in the tank of the VCO which will be discussed in the next chapter. The top view of the layout of these test inductors are shown in Fig. 3.12. The inductors are named as W8_Dout130 and W8_Dout222. The layouts are drawn in Cadence Virtuoso XL Layout editor using the foundry design kit. Note that the provisions

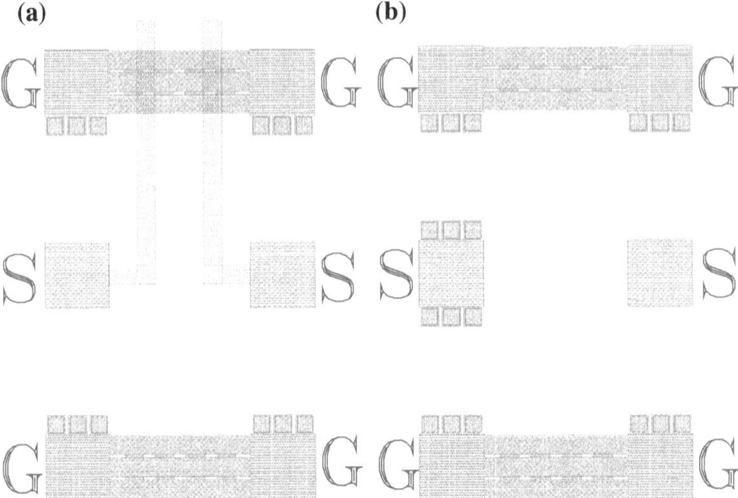

Fig. 3.13 Standard test fixtures used to characterize the parasitics: **a** Open and **b** Short

made for enabling on-wafer measurement actually take much area than the inductor itself.

The inductors will be characterized on-wafer by probing with RF probes and using a Vector Network Analyzer (VNA) to measure the Scattering or S-parameters. The RF probes used for high frequency characterization usually have the Ground Signal (GS/SG) or Ground Signal Ground (GSG) configuration. In the figure we can see the inductors placed and each port connected to the probe pads laid out in GSG configuration for a two port characterization. Using a GS/SG configuration probe one can save the area of the wafer since only two sets of contact pads are needed as compared to six for GSG. But we have chosen a GSG configuration as it gives better isolation between the signal ports. The pitch of the pads are 200 μm, drawn to match the pitch of the GSG probes. The metal turns of the MPS inductors are in M6, M5, M4, and M3. The probe pads are of size 65 × 65 μm. Figure 3.13 represents the standard 'open' and 'short' dummy test fixtures for the deembedding method which will be discussed in the next section. In the 'short' pattern of Fig. 3.13b only the signal probe pad (on the left) is shorted and the other is open. This will be used to correct the probe and pad contact parasitics.

3.6.3 Deembedding Process

For accurate on-wafer characterization of the device, the deembedding process must be done in which the parasitics due to the probe pads and the metal interconnects are subtracted from the measurement of the device under test (DUT), i.e., inductors in

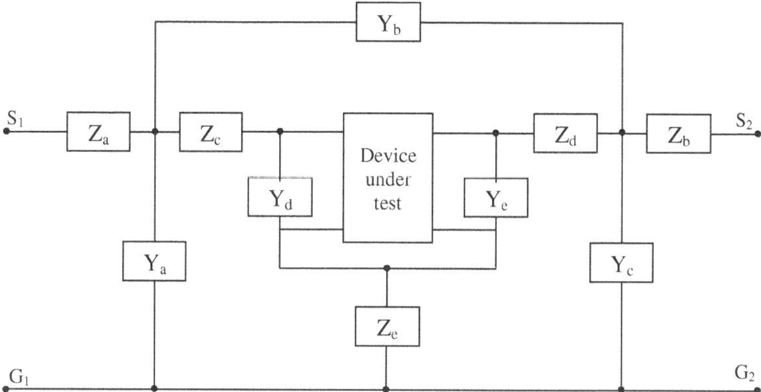

Fig. 3.14 An equivalent model of the DUT with the series and parallel parasitics originating from the probe pads and metal interconnect

this case. For this process standard test fixtures are fabricated along with the devices to measure the series and the shunt parasitics. Different deembedding processes are proposed in the literature. In Fig. 3.14 an equivalent model of the DUT with the parasitics is given [18]. The admittance Y_a, Y_b, and Y_c represents the parasitics due to the couplings between the ports. Y_d and Y_e represents coupling between the signal metal lines and the ground leads. In the case of inductor as DUT, we can see from Fig. 3.12 that the distance between the signal metal lines and the ground leads is of several microns and the coupling between them is negligible. Therefore Y_d and Y_e can be eliminated. Z_a and Z_b represent the parasitic contact resistance between the probe and the pad. Z_c and Z_d represent the interconnect parasitics. In circuits inductors can be seen connected with long interconnects and hence this can also be considered as part of the DUT and therefore Z_c and Z_d can also be removed to simplify the model [18]. Z_e represents the series impedance of the ground leads. In our test fixture design, we shielded the signal probe pad by metal (M1) layer to prevent the signal leakage to the substrate. Therefore, Z_e can also be eliminated [18, 19]. The simplified model of the DUT with the fixture parasitics is shown in Fig. 3.15. Therefore, the impedance Z_a and Z_b can be corrected with a dummy 'single short' fixture measurement, $Z_{\text{single\_short}}$, and the shunt parasitics Y_a, Y_b, and Y_c can be corrected with a dummy 'open' fixture measurement, Y_{open}. The standard 'open' and the 'single short' test fixtures are already shown in Fig. 3.13. The deembedding is therefore done in two steps as given below.

(i) The S parameters of the DUT in a fixture is measured. For this case it is the inductor as shown in Fig. 3.12. Let S_{total} represent the measured S parameters. Then S_{total} is converted to Z_{total}.

$$\left[S_{\text{total}}\right] \Rightarrow \left[Z_{\text{total}}\right] \tag{3.29}$$

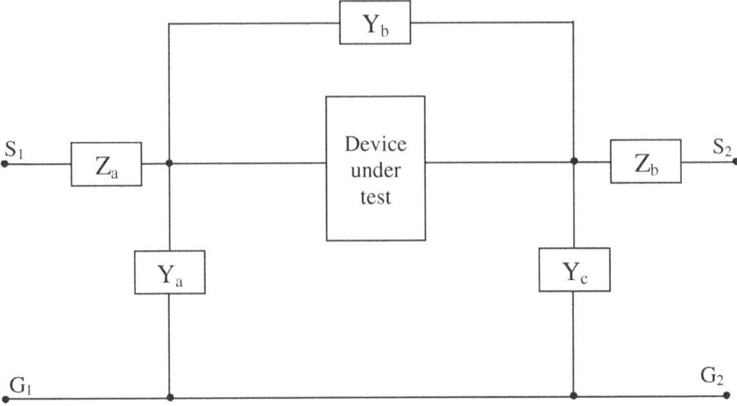

Fig. 3.15 A simplified model of the DUT with the series and parallel parasitics

The S parameters of the single short fixture is measured and let it be represented by $S_{\text{single\_short}}$. $S_{\text{single\_short}}$ is converted to $Z_{\text{single\_short}}$. Subtracting $Z_{\text{single\_short}}$ from Z_{total} will correct the effect of Z_a and Z_b and let it be represented by Z'_{DUT}.

$$\left[S_{\text{single\_short}}\right] \Rightarrow \left[Z_{\text{single\_short}}\right] \tag{3.30}$$

$$\left[Z_{\text{single\_short}}\right] = \begin{bmatrix} Z_a & 0 \\ 0 & Z_b \end{bmatrix} \tag{3.31}$$

$$\left[Z'_{\text{DUT}}\right] = \left[Z_{\text{total}}\right] - \begin{bmatrix} Z_a & 0 \\ 0 & Z_b \end{bmatrix} \tag{3.32}$$

Finally this corrected Z parameters of the DUT is converted to Y parameters.

$$\left[Z'_{\text{DUT}}\right] \Rightarrow \left[Y'_{\text{DUT}}\right] \tag{3.33}$$

(ii) The S parameters of the 'open' fixture is measured and let it be represented by S_{open}. This measurement includes the effect of contact parasitics and so $Z_{\text{single\_short}}$ must be subtracted initially. So S_{open} is converted to Z_{open} and $Z_{\text{single\_short}}$ is subtracted from Z_{open}. This will give the admittance parasitics and let it be represented by Y'_{open}.

$$\left[S_{\text{open}}\right] \Rightarrow \left[Z_{\text{open}}\right] \tag{3.34}$$

$$\left[Z'_{\text{open}}\right] = \left[Z_{\text{open}}\right] - \begin{bmatrix} Z_a & 0 \\ 0 & Z_b \end{bmatrix} \tag{3.35}$$

$$\left[Z'_{\text{open}}\right] \Rightarrow \left[Y'_{\text{open}}\right] \quad (3.36)$$

$$\left[Y'_{\text{open}}\right] = \begin{bmatrix} Y_a + Y_b & -Y_b \\ -Y_b & Y_b + Y_c \end{bmatrix} \quad (3.37)$$

The actual Y-parameters of the inductor are hence given by

$$\left[Y_{\text{DUT}}\right] = \left[Y'_{\text{DUT}}\right] - \left[Y'_{\text{open}}\right] \quad (3.38)$$

$$\left[Y_{\text{DUT}}\right] = \left[Y'_{\text{DUT}}\right] - \begin{bmatrix} Y_a + Y_b & -Y_b \\ -Y_b & Y_b + Y_c \end{bmatrix} \quad (3.39)$$

3.6.4 Measured Results and Discussion

Two of the proposed inductors were fabricated in UMC 0.18μm 1P6M RF CMOS process. The inductors as given in the previous section, have the same metal width of 8 μm, and outer diameter of 130 and 222 μm (W8_Dout130 and W8_Dout222 in Fig. 3.16. The inductors were characterized on-wafer and S parameters were measured using a Vector Network Analyzer (VNA). We have used Ground Signal Ground (GSG) RF probes. The standard de-embedding process as discussed before was performed. The results of inductance (L) and the quality factor (Q) are calculated from the two port parameters as

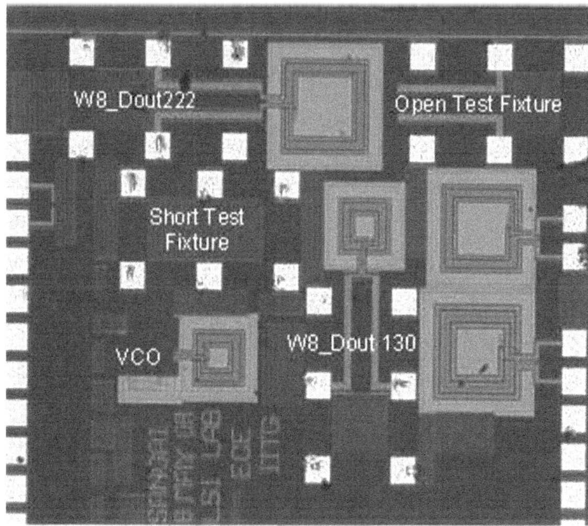

Fig. 3.16 Micrograph of the chip with inductors and the VCO

$$L = \frac{Im(1/Y_{11})}{2\pi f} \tag{3.40}$$

$$Q = \frac{Im(1/Y_{11})}{\mathrm{Re}(1/Y_{11})} \tag{3.41}$$

The measured inductance and the quality factor of the two MPS inductors of outer diameter of 130 and 222 μm are shown in Figs. 3.17 and 3.18. The inductor with the outer diameter of 130 μm has an inductance of 6.9 nH at 1 GHz with a peak quality factor of 6 at 2.1 GHz while the inductor with the outer diameter of 222 μm has an inductance of 27 nH at 1 GHz with a peak quality factor of 3 at 1.1 GHz. The measured and the simulated results agree well upto the frequency at which the quality factor peaks. Beyond this frequency the substrate loss in the inductor dominates the ohmic metal loss. The measured self resonating frequency of the MPS inductor is lower than the simulated results. This is due to the presence of a grounded guard ring which

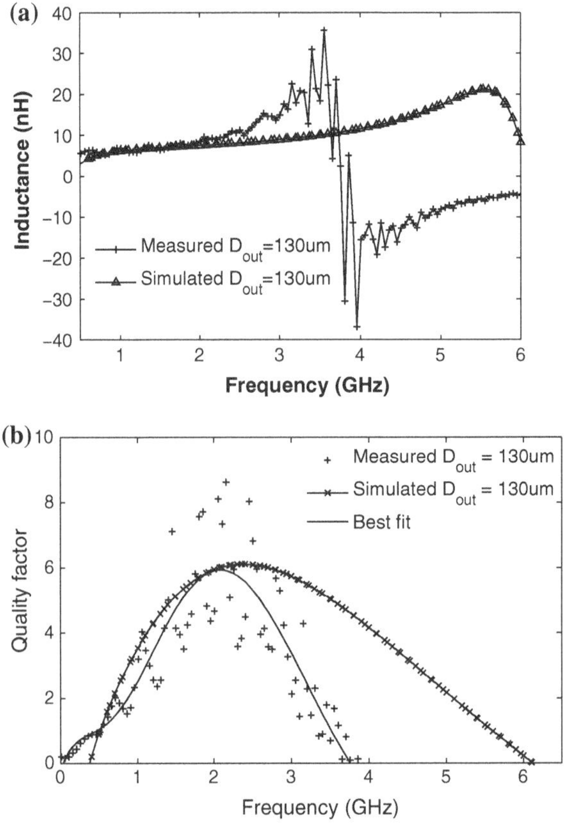

Fig. 3.17 Measured and simulated **a** Inductance and **b** Quality factor of the MPS inductors of outer diameter 130 μm

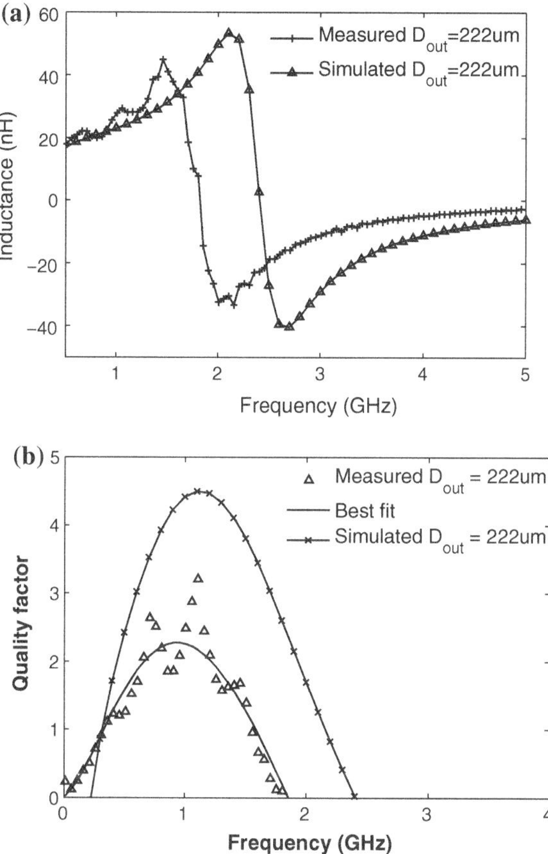

Fig. 3.18 Measured and simulated **a** Inductance and **b** Quality factor of the MPS inductors of outer diameter 222 μm

results in larger capacitance to ground. Without the guard rings in the layout of the test inductors, the self resonating frequency will increase by more than 40 % [20] and simulation will be close to the measured values. Typically, guard rings are placed around the inductor used in circuits to reduce the substrate noise coupling.

3.7 Summary

A multilayer pyramidal symmetric spiral inductor was proposed in this chapter. The proposed structure is applicable for both single ended and differential circuits. Being multilevel, the structure achieves higher inductance to area ratio and occupies smaller area as compared to its equivalent conventional planar symmetric and asymmetric pair. The performance trend of MPS inductors were demonstrated by varying its width, outer diameter and metal offsets. The symmetric nature was also illustrated with design examples. To estimate the equivalent parasitic capacitance and self resonant frequency, a compact model with a closed form expression was also developed. The results of the model were compared to that of an Electromagnetic simulator. The performance of MPS inductor was also compared to other reported symmetric inductors in literature. With multilayer pyramidal symmetric inductor, the area occupied by the inductor in integrated circuits will be reduced substantially and subsequently the cost will be minimized. Two structures of outer diameter 130 and 222 μm and width of 8 μm were fabricated and characterized. The four layer proposed structures with the outer diameter of 130 μm resulted an inductance of 6.9 nH at 1 GHz with a peak quality factor of 6 at 2.1 GHz while the inductor with the outer diameter of 222 μm has an inductance of 27 nH at 1 GHz with a peak quality factor of 3 at 1.1 GHz.

References

1. Craninckx, J., Steyaert, M.: A 1.8-GHz low-phase-noise CMOS VCO using optimized hollow spiral inductors. IEEE J. Solid-State Circuits **32**(5), 736–744 (1997)
2. Danesh, M., Long, J.R.: Differentially driven symmetric microstrip inductors. IEEE Trans. Microw. Theory Tech. **50**(1), 332–341 (2002)
3. Kuhn, W.B., Elshabini-Riad, A., Stephenson, F.W.: Centre-tapped spiral inductors for monolithic bandpass filters. Electron. Lett. **31**(8), 625–626 (1995)
4. Rabjohn, G.G.: Monolithic microwave transformers. Master's thesis, Carleton University, Ottawa, ON, Canada (1991)
5. Haobijam, G., Paily, R.: Design of multilevel pyramidically wound symmetric inductor for cmos rfics. Analog Integr. Circ. Sig. Process **63**(1), 9–21 (2010)
6. Edelstein, D.C., Burghartz, J.N.: Spiral and solenoidal inductor structures on silicon using Cu-damascene interconnects, In: Proceeding of IEEE International Interconnect Technology Conference, pp. 18–20 June 1998
7. Yim, S.-M., Chen, T., Kenneth, K.O.: The effects of a ground shield on the characteristics and performance of spiral inductors. IEEE J. Solid-State Circuits **37**(2), 237–244 (Feb. 2002)
8. Pettenpaul, E., Kapusta, H., Weisberger, A., Mampe, H., Luginsland, J., Wolff, I.: Cad models of lumped elements on gaas up to 18 ghz. IEEE Trans. Microwave Theory Tech. **36**, 294–304 (1988)
9. Zolfaghari, A., Chan, A., Razavi, B.: Stacked inductors and transformers in CMOS technology. IEEE J. Solid-State Circuits **36**(4), 620–628 (2001)
10. Tang, C.-C., Wu, C.-H., Liu, S.-I.: Miniature 3-D inductors in standard CMOS process. IEEE J. Solid-State Circuits **37**(4), 471–480 (2002)
11. Bennett, H.S., Brederlow, R., Costa, J.C., Cottrell, P.E., Huang, W.M., Immorlica, A.A., Mueller, J.E., Racanelli, M., Shichijo, H., Weitzel, C.E., Zhao, B.: Device and technology evolution for silicon-based RF integrated circuits. IEEE Trans. Electron Devices **52**(7), 1235–1258 (2005)
12. Farina, M., Rozzi, T.: A 3-D integral equation-based approach to the analysis of real-life MMICs-application to microelectromechanical systems. IEEE Trans. Microw. Theory Tech. **49**(12), 2235–2240 (2001)
13. Koutsoyannopoulos, Y.K., Papananos, Y.: Systematic analysis and modeling of integrated inductors and transformers in RFIC design. IEEE Trans. Circuits Syst. II **47**(8), 699–713 (2000)

14. Long, J.R., Copeland, M.A.: The modeling, characterization, and design of monolithic inductors for silicon RF IC's. IEEE J. Solid-State Circuits **32**(3), 357–369 (1997)
15. Haobijam, G., Paily, R.: Performance study of fixed value inductors and their optimization using electromagnetic simulator. Microwave Opt. Technol. Lett. **50**(5), 1205–1210 (2008)
16. Teo, T.H., Choi, Y.-B., Liao, H., Xiong, Y.-Z., Fu, J.S.: Characterization of symmetrical spiral inductor in 0.35 μm CMOS technology for RF application. Solid State Electron. **48**(9), 1643–1650 (2004)
17. EUROPRACTICE IC Service. http://www.europractice-ic.com
18. Aguilera, J., Berenguer, R.: Design and Test of Integrated Inductors for RF Applications. Kluwer Academic Publishers, Boston (2003)
19. Aktas, A., Ismail, M.: Pad de-embedding in RF CMOS. IEEE Circuits Devices Mag. **17**(3), 8–11 (2001)
20. Pun, A.L.L., Yeung, T., Lau, J., Clement, J.R., Su, D.K.: Substrate noise coupling through planar spiral inductor. IEEE J. Solid-State Circuits **33**(6), 877–884 (1998)

Chapter 4
Implementation of the MPS in Voltage Controlled Oscillator

4.1 Introduction

Voltage controlled oscillators are used in many analog and RF signal processing systems. It is one of the key building blocks of RF transceivers. With the rapid growth and advancement of wireless communication systems and standards, there has been an increasing demand of high performance and fully integrated GHz voltage controlled oscillators. VCO's may be implemented as ring oscillators, relaxation oscillators or tuned oscillators. Ring oscillators and relaxation oscillators are easily integrable and suitable for low power but have poor phase noise performance. On the other hand phase shift or RC oscillators are stable and provide a well-shaped sine wave output. However, RC oscillators are restricted to high frequency applications because of various reasons. RC oscillator encounters high phase noise at high frequencies which results in instability of frequency. It requires very small resistor value (typically in one-tenth of Ohms) at high frequencies, which is difficult to realise on-chip. At high frequencies RC Oscillator require very high gain transistors because of losses encountered in RC network and are limited by their banwidth constraints to produce the desired phase shift for oscillation. The frequency of oscillation is proportional to $1/2\pi RC\sqrt{2N}$ where N is the number of stages. At high frequencies, large number of RC sections is required to have frequency stability with substantial phase noise. Hence, RC oscillators are good for frequencies up to 1 MHz. Thus the best choice for high frequency voltage controlled oscillators are LC oscillator with less phase noise and realizable L and C component values onchip. LC oscillators have low phase noise and jitter at high frequencies as compared to RC oscillators.

Wireless applications require a low phase noise and therefore LC oscillators are preferred. Of the various LC oscillator topologies, the cross-coupled differential LC oscillator topology is the optimum choice. The differential output also eliminates the need of single ended to differential conversion circuitry. In differential VCO implementation the inductor of the LC tank is implemented with a pair of planar spiral inductors by connecting their inner loops in series. Since the currents always flow in opposite direction in these two inductors, there must be enough spacing between

G. Haobijam and R. P. Palathinkal, *Design and Analysis of Spiral Inductors*,
DOI: 10.1007/978-81-322-1515-8_4, © Springer India 2014

them to minimize electromagnetic coupling. As a result, the overall area occupied by the inductors is very large. To eliminate the use of two inductors and reduce the chip area consumption, the center tapped spiral inductor [1] or a symmetrical inductor [2] can be used. This type of winding of the metal trace was first applied to monolithic transformers [3]. The symmetric inductor is realized by joining groups of coupled microstrip from one side of an axis of symmetry to the other using a number of cross-over and cross-under connections. Symmetrical inductors under differential excitation results in a higher quality factor and self resonance frequency and occupies less area than its equivalent asymmetrical inductors. In this chapter, the implementation of the multilayer pyramidal symmetric inductor in a 2.5 GHz voltage controlled oscillator is presented. In Sect. 4.2 the design of the passive elements of the tank circuit of the VCO are explained. In Sect. 4.3 the design of the 2.4 GHz VCO circuit is discussed. In Sect. 4.4 the simulation results are reported and in Sect. 4.5 the measurement results are discussed. Finally the chapter is summarized in Sect. 4.6.

4.2 Passive Elements of the LC Tank

In the following subsections, the design of the inductor and the varactor of the LC tank circuit are discussed.

4.2.1 Inductor Design

On-chip inductors fabricated on Silicon substrate suffers from poor quality factor due to ohmic and substrate losses. However for a chosen technology and desired frequency of the application, the layout design parameters of the on-chip inductor can be optimized for best quality factor and minimum area as discussed in Chap. 2. In an LC oscillator the most critical circuit element is the inductor. For a low phase noise oscillator the quality factor of the LC tank must be sufficiently high. The quality factor of the LC tank is dominated by the quality factor of the inductor. Hence an inductor with a high quality factor must be designed. Also the inductance value must be chosen such that it satisfies the tank amplitude and the oscillator startup constraints for the maximum bias current allowed by the design specifications. To minimize the area, the inductor of the LC tank is implemented with the new multilayer pyramidal symmetric inductor structure. The VCO was designed with minimum VCO core power constraint of 5 mW. So, this gives the design constraint on the maximum bias current for a supply voltage of 1.8 V.

$$\begin{aligned} I_{\text{bias}} &\leq I_{\text{max}} \\ I_{\text{bias}} &\leq 2.7\,\text{mA} \end{aligned} \tag{4.1}$$

Now considering a minimum tank amplitude of 1 V, we get the minimum parallel resistance requirement of the tank given by

$$I_{\text{bias}} \times R_p \geq V_{\text{tank,min}} \tag{4.2}$$

This implies

$$R_p \geq \frac{V_{\text{tank,min}}}{I_{\text{bias}}} \tag{4.3}$$

Therefore, choosing a bias current of 2 mA we get a minimum R_p of 500 Ω. Now, we can find a suitable inductor with high quality factor and $R_p \geq 500\,\Omega$. We know that the quality factor of the lossy tank is given by

$$\frac{1}{Q_{\text{tank}}} = \frac{1}{Q_L} + \frac{1}{Q_C} \tag{4.4}$$

The quality factor of the tank will be dominated by the quality factor of the inductor since the quality factor of the capacitor is very high as compared to that of the inductors. Therefore,

$$\frac{1}{Q_{\text{tank}}} = \frac{1}{Q_L} \tag{4.5}$$

Then,

$$\begin{aligned} Q_{\text{tank}} &= Q_L \\ \frac{R_p}{\omega_o L} &= \frac{\omega_o L}{R_s} \end{aligned} \tag{4.6}$$

This implies,

$$\begin{aligned} R_p &= \frac{(\omega_o L)^2}{R_s} \\ &= (\frac{\omega_o L}{R_s})^2 R_s \\ &= Q^2 R_s \end{aligned} \tag{4.7}$$

where R_s is the series resistance of the inductor. A number of MPS inductors were simulated. The MPS inductor layout parameters were varied. The metal width and the diameter were adjusted so that the inductor quality factor peaks around 2.4 GHz. Many structures were simulated as discussed in the performance study of MPS inductors in the previous chapter. The inductor which has higher quality factor, small series resistance and $R_p \geq 500\,\Omega$ is selected. An MPS inductor which has an inductance of 8 nH at 2.4 GHz was chosen. It has an outer diameter of 130 μm, inner diameter of 54 μm, metal width of 8 μm and offset of 2 μm was chosen. The

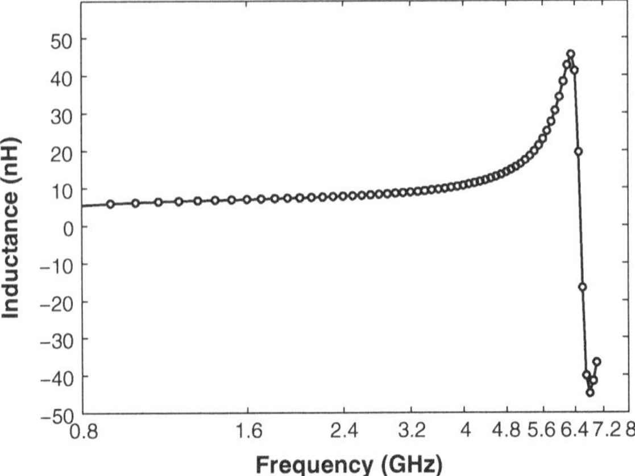

Fig. 4.1 Inductance plot of the tank MPS inductor

inductance and the quality factor under differential excitation is shown in Figs. 4.1 and 4.2. The inductor has a peak quality factor of 8 under differential excitation and inductance of 8 nH. At 2.4 GHz the quality factor is around 6 and the series resistance is 22.5 Ω. This results in a parallel resistance of 810 Ω. The structure is simulated using an EM simulator by defining all the process parameters according to the UMC foundry design kit for the chosen 0.18 μm RF CMOS process. The structure was simulated from 0 to 10 GHz. This inductor resonates at around 6.5 GHz as can be seen from the figure. The equivalent π model parameters extracted at 2.4 GHz is shown in Fig. 4.3.

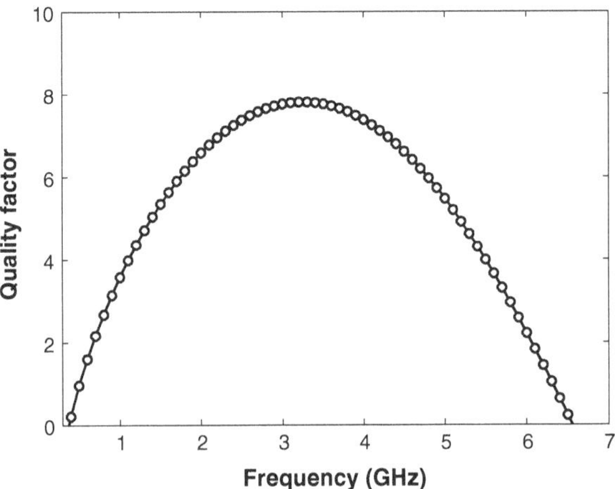

Fig. 4.2 Quality factor plot of the tank MPS inductor

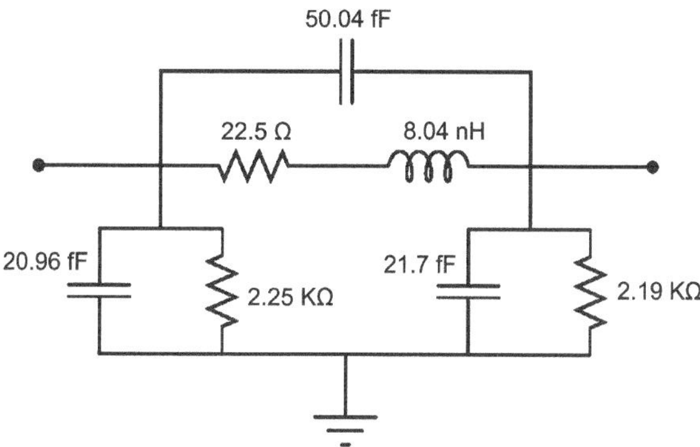

Fig. 4.3 Parameters of the π model of the tank MPS inductor extracted at 2.4 GHz

4.2.2 Varactor Design

The total capacitor of the LC tank includes the combination of the tank capacitor connected across the inductor, the NMOS and PMOS parasitic capacitances and the inductor's parasitic capacitance. The details of the VCO circuit is presented later in next section. The oscillator frequency will be tuned by varying capacitance of the tank with a controlled voltage. This varactor can be implemented with a varactor diode or a MOS varactor. Studies on the use of MOS varactor and varactor diode have shown that the performance of both MOS varactor VCO's is superior to that of the diode varactor VCO [4]. In this work the tank varactor is implemented using a MOS capacitor. The capacitor is formed by the polysilicon gate and the channel of a MOSFET. The capacitance of this MOS device varies non-linearly as the DC gate bias of the MOSFET is varied through accumulation, depletion and inversion. Therefore, this structure which is always present in a CMOS process is used as the tuning element of an oscillator. Both NMOS and PMOS varactors are possible but PMOS varactors are preferred since NMOS varactors are more sensitive to substrate-induced noise, as it cannot be implemented in a separate p-well. Different variations in the MOS capacitors are explored in the literature. In this work a PMOS inversion mode varactor is implemented. The drain and source of the PMOS is connected together to form one terminal of the capacitor and the gate forms the other terminal. The bulk is connected to the highest positive voltage available in the circuit i.e the power supply V_{dd}. The tuning range of the PMOS capacitor with this connection is much wider than for the PMOS capacitor with bulk, drain and source connected together, since the former capacitor is working in the strong, moderate, or weak inversion region only, and never enters the accumulation region [4]. The maximum capacitance will be given by C_{ox} where

$$C_{ox} = \frac{3.9\varepsilon_0 WL}{t_{ox}} \tag{4.8}$$

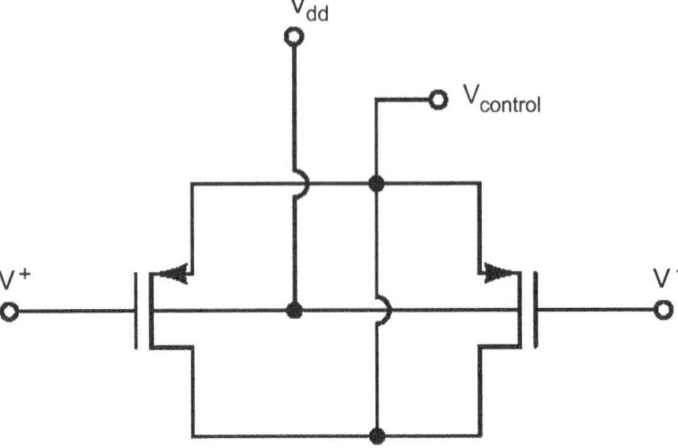

Fig. 4.4 Schematic of the PMOS varactor

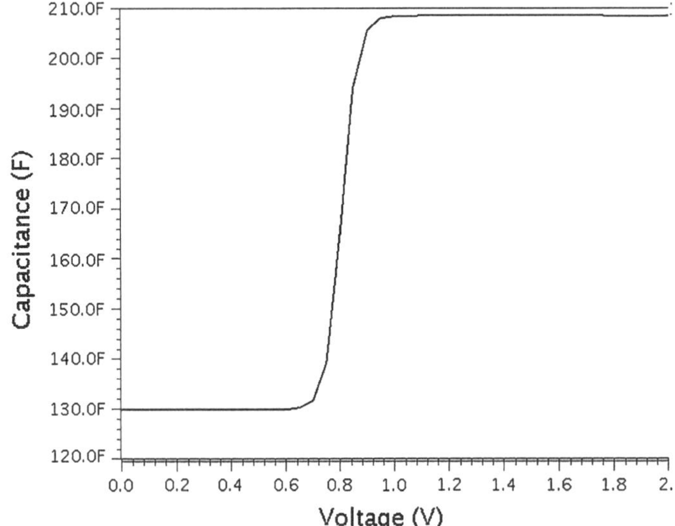

Fig. 4.5 Variation of the total tank capacitance with the control voltage

where t_{ox} is the thickness of the gate oxide. In this process the gate oxide thickness is approximately 42 Å and the capacitance is approximately 822×10^{-5} pF/μm^2. For an oscillation of 2.4 GHz and a chosen inductance of 8 nH the required total tank capacitance is 0.55 pF. Since the parasitic capacitance will also contribute to the tank capacitance, the tank C will be less than 0.55 pF. Considering the parasitics the width of the PMOS capacitor was adjusted so that the circuit oscillates at 2.4 GHz. The tank capacitor was implemented in a differential manner by a series connection of two inversion mode PMOS transistors as shown in Fig. 4.4. Each PMOS capacitor has 18 fingers and each finger of width 10 μm and with minimum gate length of 0.18μm. Therefore each capacitor of the tank will have an overall width of 180 μm. So, the

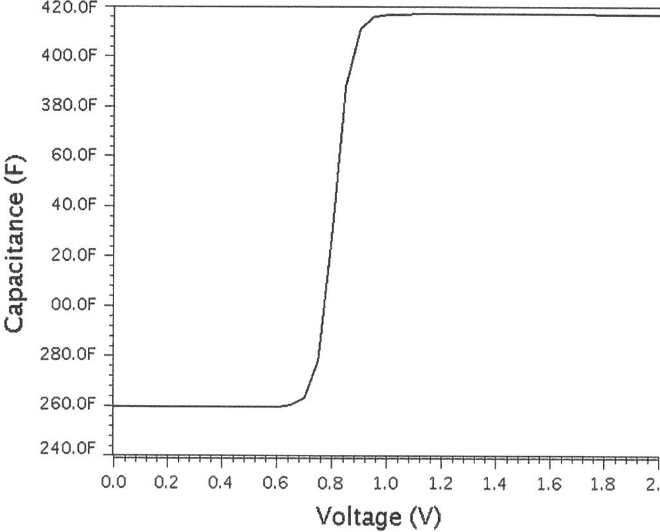

Fig. 4.6 Variation of the capacitance for a single PMOS of the tank capacitor with the control voltage

total tank C is the effective capacitance of the series connected PMOS transistors. The variation of the total tank capacitance with the control voltage connected to drain and source of each tank capacitor is shown in Fig. 4.5. Each tank C varies from 260 fF to 416 fF as can be seen from Fig. 4.6 and therefore the overall tank C varies from 130 to 208 fF.

4.3 VCO Circuit Design

The cross-coupled differential LC oscillator topology using both PMOS and NMOS is shown in Fig. 4.7. The oscillation amplitude of this topology is larger and the phase noise is lower than the oscillators using only NMOS. The rise and fall time symmetry

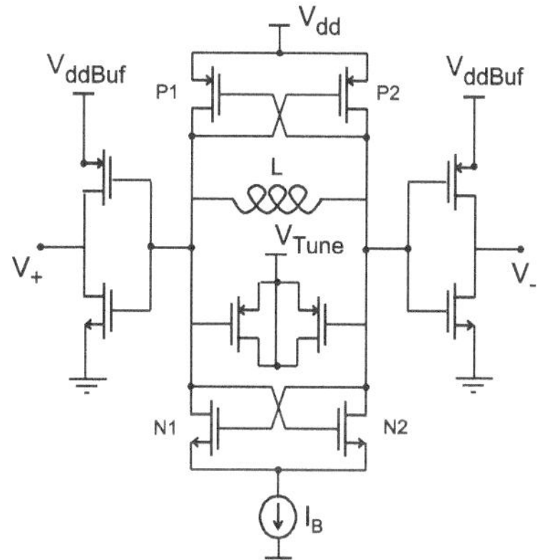

Fig. 4.7 Schematic of the cross coupled LC voltage controlled oscillator

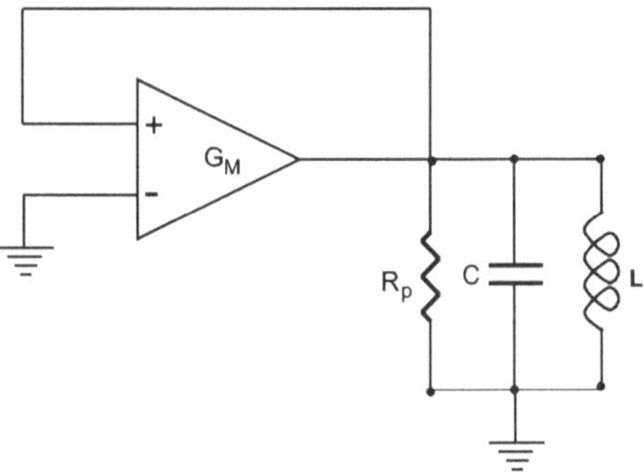

Fig. 4.8 Transconductance model of the oscillator

is also better resulting in smaller $1/f^3$ conversion [5]. In an LC oscillator the most critical circuit element is the inductor. For a low phase noise oscillator the quality factor of the LC tank must be sufficiently high. The quality factor of planar inductors will be higher than multilayer inductors. However multilayer inductors occupy less than 50 % of the area for the same inductance [6]. Performance also needs to be traded off with the cost and it would be advantageous to use multilayer inductors as long as the design specifications are satisfied. The inductance value was chosen such that it satisfies the tank amplitude and oscillator startup constraints for the maximum bias current allowed by the design specifications [7].

The differential oscillator of Fig. 4.7 can be viewed as a negative resistance LC oscillator as shown in Fig. 4.8. The active device is a simple transconductance (G_M) amplifier connected in positive feedback to an LC tank circuit. The tank circuit sees a negative resistance of $\frac{-1}{G_M}$ looking back into the transconductor output. As per the Barkhausen criteria for the circuit to oscillate, this negative resistance will exactly cancel the equivalent parallel resistance of the tank circuit. In other words, the active device must add enough energy to the circuit to cancel the total losses of the tank circuit. For the cross coupled complementary differential oscillator the negative resistance seen across the tank is

$$R_{\text{negative}} = \frac{-2}{G_{M(\text{NMOS})}} + G_{M(\text{PMOS})} \tag{4.9}$$

The oscillation condition requires that the closed loop gain be of atleast unity magnitude and zero phase angle. Therefore,

$$\left| \frac{-2}{G_{M(\text{NMOS})}} + G_{M(\text{PMOS})} \right| = R_p \tag{4.10}$$

where R_p is the equivalent parallel resistance of the tank circuit. The ratio of parallel resistance to negative resistance is called the safety start up factor, which is generally chosen to be greater than 3. With this condition the minimum size of the PMOS and the NMOS transistors is determined. For this design the chosen MPS tank inductor has a parallel resistance of 810 Ω. So $G_{M(\mathrm{NMOS})} + G_{M(\mathrm{PMOS})} = 10\,\mathrm{mS}$ was chosen such that the oscillation safety start up factor is 4. Because G_M is proportional to $\frac{W}{L}$, the device width can be minimized by using the smallest allowable gate length. The MOS devices were implemented using the minimum gate length allowed in 0.18 μm process. This minimizes the gate area and thus the gate capacitance. In order to set $G_{M(\mathrm{NMOS})} = G_{M(\mathrm{PMOS})}$, the PMOS devices must be approximately twice the size of the NMOS devices. The transistor width can be estimated using the following device equation.

$$I_{\mathrm{Dsat}} = \frac{k_p}{2}\frac{W}{L}(V_{gs} - V_t)^2 \tag{4.11}$$

$$g_m = \frac{dI_{\mathrm{Dsat}}}{dV_{gs}} = k_p\frac{W}{L}(V_{gs} - V_t) \tag{4.12}$$

After a number of iterations and using the foundry model parameters the width for the PMOS and NMOS devices were chosen to be 38 and 15 μm respectively. The tail current control device was implemented as a simple NMOS current mirror. This tail current device can alter the voltage swing across the tank circuit of the oscillator. The negative resistance seen across the tank circuit can be varied by changing this current and therefore the actual equivalent parallel resistance (R_p) of the resonator can be experimentally determined by finding the lowest bias current at which the circuit will oscillate. To measure the oscillator output using 50 Ω test equipment, the output of the oscillator is connected to a buffer. The size of the buffer transistors are adjusted so that it can drive the 50 Ω test equipment.

4.4 VCO Simulation

The VCO circuit is simulated using Spectre of Cadence. The simulations are performed using the UMC foundry design kit of 0.18 μ 1P6M RF CMOS process technology. The tank inductor is simulated using an EM simulator and the π model of the multilayer pyramidal symmetric inductor extracted at 2.4 GHz is used in the circuit simulation. In order to trigger the oscillation, a short current pulse is generated from a piece-wise-linear current source connected in parallel with the tank circuit. Figure 4.9 shows the steady state output of the oscillator. The single ended and the differential output are shown. Figure 4.10 shows the tuning curve of the oscillator. The tuning range is from 2.33 to 2.51 GHz for the tuning voltage varying from 0.7 to 1.8 V. This results in a bandwidth of 180 MHz and a gain, K_{vco} of around 163 MHz/V. The output power spectrum is shown in Fig. 4.11.

The most important design constraint of the VCO is the phase noise. Phase noise is essentially a random deviation in frequency which can also be viewed as a random

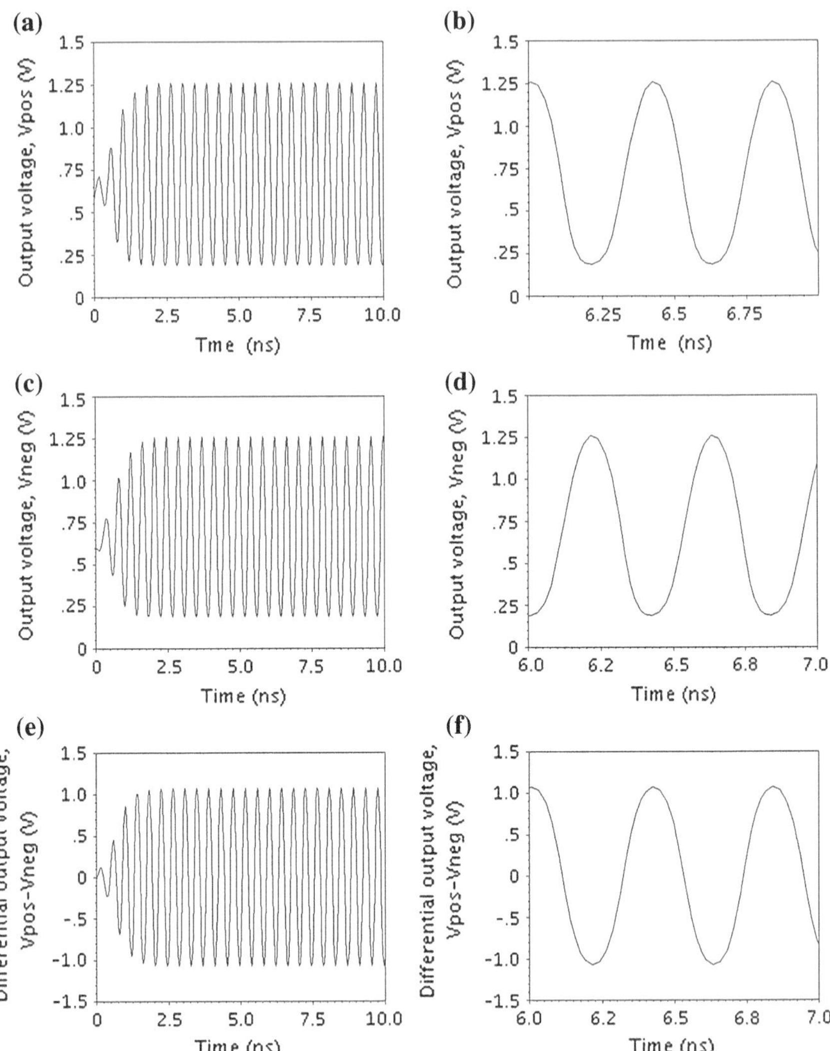

Fig. 4.9 **a** Single ended outputs at positive node of the VCO. **b** Enlarged version of figure (**a**). **c** Single ended outputs at negative node of the VCO. **d** Enlarged version of figure (**c**). **e** Differential output of the VCO and, **f** Enlarged version of (**e**)

variation in the zero crossing points of the time-dependent oscillator waveform. A real oscillator is described as

$$V_{\text{out}}(t) = V_o(t) \times y[2\pi f_c t + \phi(t)] \tag{4.13}$$

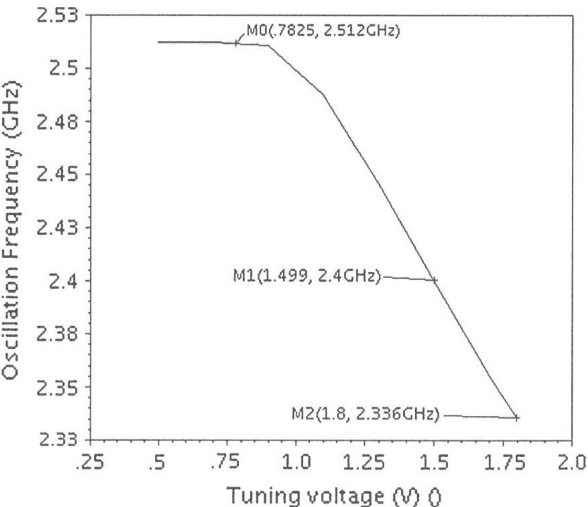

Fig. 4.10 Simulated tuning characteristic of the VCO

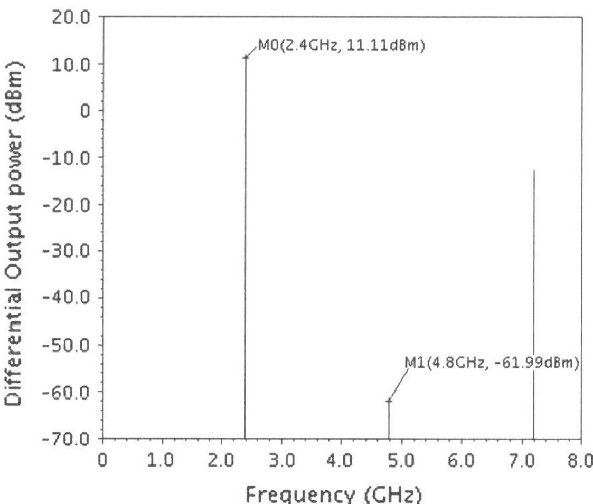

Fig. 4.11 Simulated output power spectrum of the VCO

where y is a periodic function, V_o is the constant amplitude, f_c is the center frequency, ϕ is the fixed phase of the oscillator. The fluctuations introduced by $V_o(t)$ and $\phi(t)$ will result in sidebands close to f_c with symmetrical distribution around f_c. This frequency fluctuations are characterized by the single sideband noise spectral density normalized to the carrier signal power. It is given by

$$Ł_{\text{total}}(f_c, \Delta f) = 10\log\left[\frac{P_{\text{sideband}}(f_c + \Delta f, 1\text{Hz})}{P_{\text{carrier}}}\right] \tag{4.14}$$

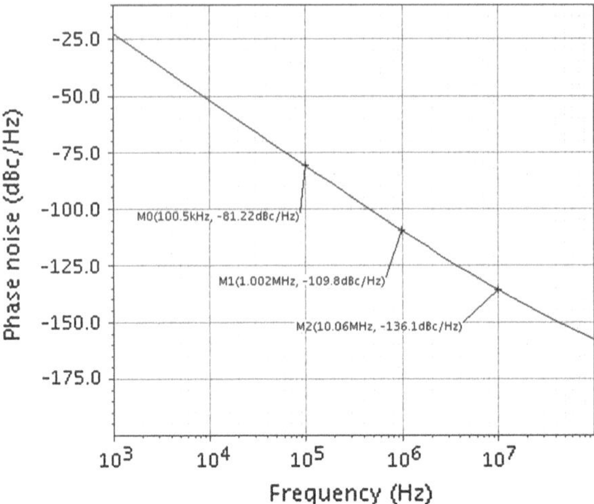

Fig. 4.12 Simulated phase noise of the VCO

where P_{carrier} is the carrier signal power at frequency f_c and $P_{\text{sideband}}(f_c+\Delta f, 1\ Hz)$ is the single sideband power at the offset of Δf from the carrier f_c at a measurement bandwidth of 1 Hz. It has units of decibels below the carrier per hertz (dBc/Hz). The total phase noise includes both the amplitude and the phase fluctuations but it is dominated by the phase part of the phase noise. Phase noise is also simulated by performing a steady state noise analysis in Spectre. All the parasitic capacitance due to the bondpads, the parasitic bondwires inductance and the electrostatic discharge protection circuits were included in the simulation. This results in a phase noise of -109 dBc/Hz at an offset frequency of 1 MHz. The phase noise is plotted in Fig. 4.12. The layout of the VCO is shown in Fig. 4.13. The chip area of the VCO without the pads is $193 \times 300\,\mu\text{m}$ only.

4.5 Measurement Results and Discussion

The VCO was fabricated in UMC 0.18 μm 1P6M MM/RFCMOS process. The VCO measurement was done on a prototype board developed using the QFN packaged chip and the RF outputs were available through standard SMA connector. Agilent E4407B ESA-E Series, Spectrum Analyzer was used for the measurements. The micrograph of the chip was shown in Fig. 3.16 in the previous chapter. The prototype board and the measurement set up is shown in Fig. 4.14. Figure 4.15 shows the measured output power spectrum for a carrier frequency of 2.54 GHz. The single ended output power is -10 dBm. A span of 200 MHz and a resolution bandwidth of 1 MHz are chosen for this measurement. The phase noise at different offsets from the carrier is shown in Fig. 4.16. The phase noise is -98, -108 and -128 dBc/Hz at an offset frequency

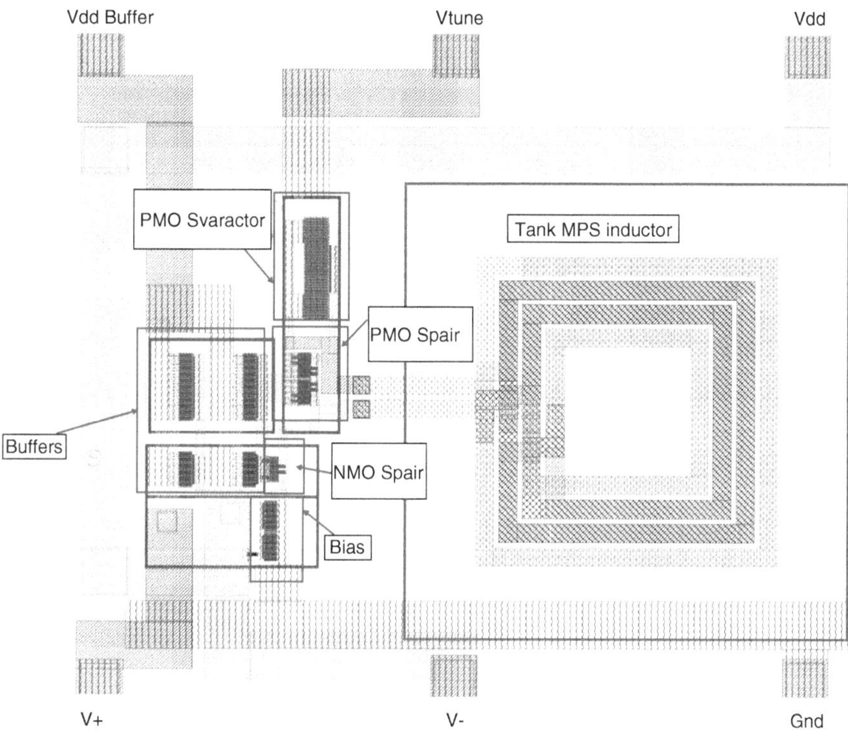

Fig. 4.13 Layout of the VCO

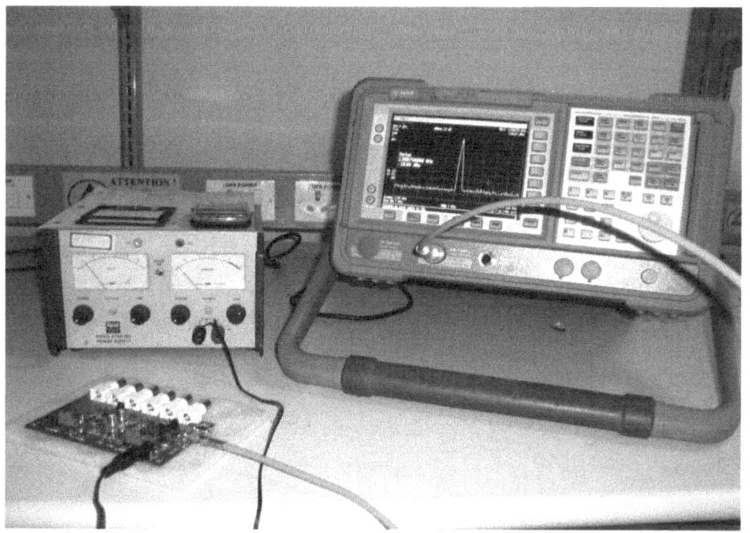

Fig. 4.14 Testing and measurement of the VCO on the prototype board

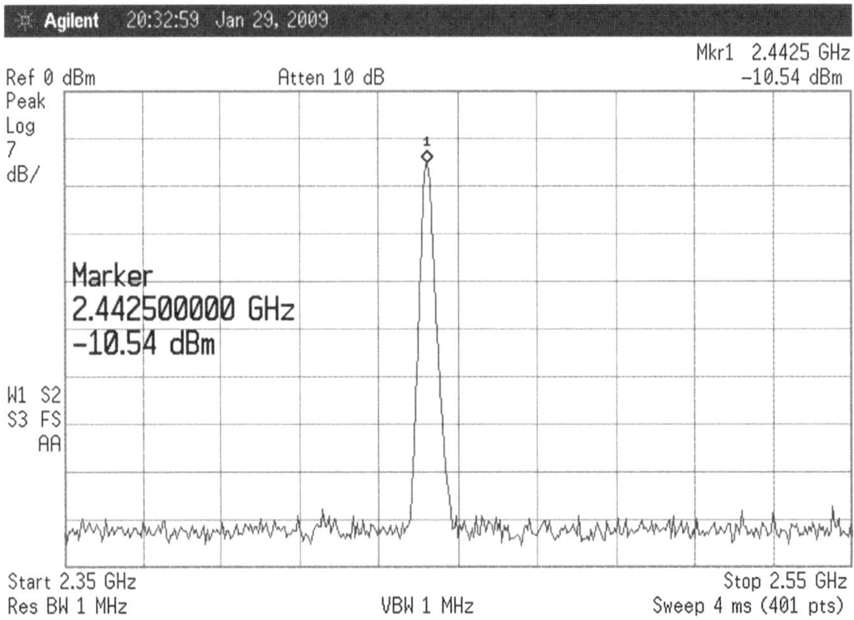

Fig. 4.15 Measured output power spectrum of the VCO

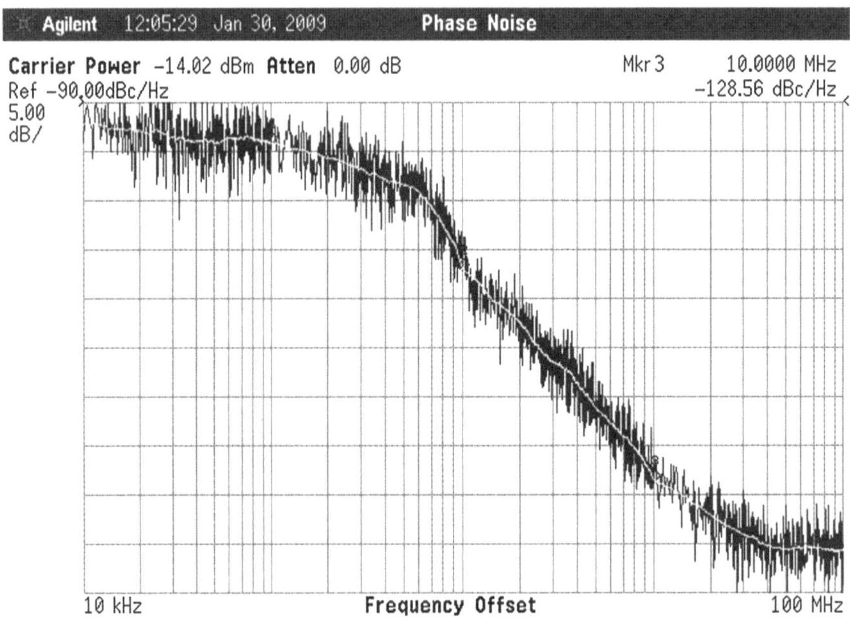

Fig. 4.16 Measured phase noise of the VCO

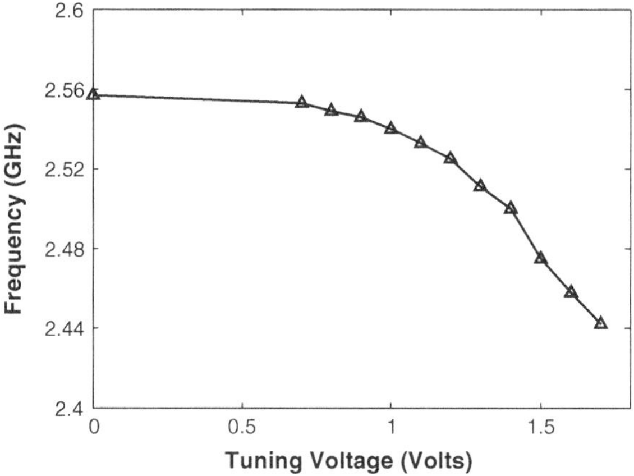

Fig. 4.17 Measured tuning characteristics of the VCO

of 100 KHz, 1 and 10 MHz respectively from the carrier. It may be noted that if differential output was measured, the phase noise will be lowered by 6 dBc since the oscillation amplitude would almost double. The supply voltage is 1.8 V and the VCO core consumes a power of 5 mW. Figure 4.17 shows the tuning characteristic of the VCO. The measured tuning frequency range is 2.441–2.557 GHz for a control voltage ranging from 0 to 1.8 V. This corresponds to a tuning of 116 MHz bandwidth and gain K_{VCO} of 68.23 MHz/V. The performance of the VCO with the new multilayer inductor can be compared to other oscillators based on the widely used figure of merit (FOM) [8]. At carrier frequency of 2.545 GHz the FOM is 180. The performance is summarized in Table 4.1. The performance is comparable to other oscillators of [9, 10] with multilayer tank inductor. The performance of the VCO meets the specifications for various applications in the 2.4 to 2.5 GHz unlicensed ISM band and with the new inductor it would be advantageous to have a large reduction in the chip area.

Table 4.1 VCO performance summary

Parameter	Values
Supply voltage	1.8 V
Current (core)	5 mA
Tuning range	2.441–2.557 GHz
Tuning bandwidth	116 MHz
Output power (50 Ω load)	−10 dBm
Phase noise at 100 KHz offset	−98 dBc/Hz
Phase noise at 1 MHz offset	−108 dBc/Hz
Phase noise at 10 MHz offset	−128 dBc/Hz

4.6 Summary

In this chapter the design and implementation of an integrated cross coupled LC voltage controlled oscillator using the MPS inductor in 0.18 μm RF CMOS technology was presented. The cross coupled topology results in higher oscillation amplitude and also reduces the 1/f noise upconversion. The VCO circuit was simulated and verified using Cadence Custom IC design tools. The oscillator attains a steady state in less than 2.5 ns. With MPS inductor as tank inductor the area of the VCO chip will be reduced. Employing the new inductor in the tank, a satisfactory performance of the VCO with a phase noise −98 dBc/Hz at 100 KHz offset is achieved. With smaller area and low parasitics, the new inductor can be used to reduce the cost of RFIC's.

References

1. Kuhn, W.B., Elshabini-Riad, A., Stephenson, F.W.: Centre-tapped spiral inductors for monolithic bandpass filters. Electron. Lett. **31**(8), 625–626 (1995)
2. Danesh, M., Long, J.R.: Differentially driven symmetric microstrip inductors. IEEE Trans. Microw. Theory Tech. **50**(1), 332–341 (2002)
3. Rabjohn, G.G.: Monolithic microwave transformers, master's thesis. Carleton University, Ottawa (1991)
4. Andreani, P., Mattisson, S.: On the use ofMOS varactors in RF VCOs. IEEE J. Solid State Circ. **35**(6), 905–910 (2000)
5. Hajimiri, A., Lee, T.H.: Design issues in CMOS differential LC oscillators. IEEE J. Solid State Circ. **34**(5), 717–724 (1999)
6. Zolfaghari, A., Chan, A., Razavi, B.: Stacked inductors and transformers in CMOS technology. IEEE J. Solid State Circ. **36**(4), 620–628 (2001)
7. Craninckx, J., Steyaert, M.: A 1.8-GHz low-phase-noise CMOS VCO using optimized hollow spiral inductors. IEEE J. Solid State Circ. **32**(5), 736–744 (1997)
8. Razavi, B.: RF Microelectronics. Prentice Hall, Upper Saddle River (1998)
9. Lee, S-H., Chuang, Y-H., Chi, L-R., Jang, S-L., Lee, J-F.: A low-voltage 2.4 GHz VCO with 3D helical inductors. In: Proceedings of IEEE Asia Pacific Conference on Circuits and Systems, pp. 518–521 (2006)
10. Lam, C., Razavi, B.: A 2.6 GHz/5.2 GHz CMOS voltage-controlled oscillator. In: Proceedings of IEEE International Solid-State Circuits Conference, pp. 402–403 (1999)

Erratum to: Design and Analysis of Spiral Inductors

Genemala Haobijam and Roy Paily Palathinkal

Erratum to:
G. Haobijam and R. P. Palathinkal, *Design and Analysis of Spiral Inductors*, DOI 10.1007/978-81-322-1515-8

Affiliation of the author "Dr. Haobijam" was published incorrectly. It should read as "School of Computing and Electrical Engineering, Indian Institute of Technology Mandi, Mandi, Himachal Pradesh, India" instead of "R&D 3, ATG, SEL, Samsung Noida Mobile Centre, Noida, India".

The online version of the original book can be found at 10.1007/978-81-322-1515-8.

G. Haobijam (✉)
School of Computing and Electrical Engineering, Indian Institute of Technology Mandi, Mandi, Himachal Pradesh, India
e-mail: genemala@gmail.com

R. P. Palathinkal
Department of Electronics and Electrical Engineering, Indian Institute of Technology Guwahati, Guwahati, Assam, India
e-mail: roypaily@iitg.ernet.in

DOI: 10.1007/978-81-322-1515-8_5,

About the Author

Dr. Genemala Haobijam holds a Ph.D. in Electronics and Communication Engineering from Indian Institute of Technology Guwahati, India; M.Tech. degree in Microelectronics and VLSI Design from Shri Govindram Seksaria Institute of Technology and Science, Indore; and the B.E. degree in Electronics and Telecommunication Engineering from Shri Sant Gajanan Maharaj College of Engineering, Maharashtra. She was born in Imphal, Manipur, India. She has served as Assistant Professor at Indian Institute of Technology Mandi, Himachal Pradesh between 2010 and 2012. She has taught courses in Analog/Digital Circuits and System Design. She was awarded twice and given recognition for her teaching at Indian Institute of Technology Mandi. Her area of research and teaching interest focuses on analysis and design of Radio Frequency Integrated Circuits and Analog & Mixed Signal Circuits. She won the first prize in Design Contest of 22nd International Conference on VLSI Design, January 2009. She has published several papers in peer reviewed international journals and conferences. She has also worked for IBM India Pvt. Ltd. between 2009 and 2010. She is currently working at Samsung India Electronics Pvt. Ltd. She is a member of IEEE, IEEE Solid State Circuits Society, IEEE Women in Engineering.

Prof. Roy Paily received his B.Tech. degree in Electronics and Communication Engineering from College of Engineering, Trivandrum, India in 1990. He obtained the M.Tech. and Ph.D. from Indian Institute of Technology Kanpur and Indian Institute of Technology Madras in 1996 and 2004 respectively, in the area of Semiconductor Devices. Presently, he is working as a Professor in the Department of Electronics and Electrical Engineering, Indian Institute of Technology Guwahati. Before joining academia, he worked as a Process Engineer in a hard disk manufacturing company. His research interests are VLSI devices/circuits and MEMS, and has published about 100 articles in journals and conferences. His team has taped out two silicon CMOS chips though Europractice IC Service under a Research and Development project "Special Manpower Development Project in VLSI Design and Related Software" sponsored by DIT, India. He has developed a web-based course, "Integrated Circuits Technology" under National program on Technology Enhanced Learning. "Digital VLSI Design Virtual Lab" is another online course developed through the initiative of MHRD. He is a member of IEEE and VLSI Society of India and a life member of ISSS and IETE.

G. Haobijam and R. P. Palathinkal, *Design and Analysis of Spiral Inductors*,
DOI: 10.1007/978-81-322-1515-8, © Springer India 2014

Index

G. Haobijam and R. P. Palathinkal, *Design and Analysis of Spiral Inductors*,
DOI: 10.1007/978-81-322-1515-8, © Springer India 2014

Zeitfracht Medien GmbH
Ferdinand-Jühlke-Straße 7
99095 Erfurt, Deutschland
produktsicherheit@kolibri360.de